ELECTROSPINNING OF NANOFIBERS IN TEXTILES

ELECTROSPINNING OF NANOFIBERS IN TEXTILES

A. K. Haghi, PhD
Professor, University of Guilan, Iran;
Editor-in-Chief, International Journal of
Chemoinformatics and Chemical Engineering

Apple Academic Press
TORONTO NEW JERSEY

International Standard Book Number: 978-1-926895-04-8 (Hardback)

Printed in the United States of America

10 9 8 7 6 5 4 3 2 1

Library and Archives Canada Cataloguing in Publication

Electrospinning of nanofibers in textiles/by A.K. Haghi.

Includes bibliographical references and index.
ISBN 978-1-926895-04-8
1. Electrospinning. 2. Nanofibers. 3. Textile fabrics. Title.

TA418.9.F5H34 2011 677›.02832 C2011-906631-9

Apple Academic Press also publishes its books in a variety of electronic formats. Some content that appears in print may not be available in electronic format. For information about Apple Academic Press products, visit our website at **www.appleacademicpress.com**

Contents

List of Abbreviations

AFM	Atomic force microscopy
CV	Coefficient of variation
dB	Decibels
DMF	Dimethylformamide
EOS	Equivalent opening size
FCCD	Face-centered central composite design
MAT	Medial axis transformation
MFD	Mean fiber diameter
ODF	Orientation distribution function
PAN	Polyacrylonitrile
PDLA	Poly D, L-lactide
PEOT/PBT	Polyethylene oxide terephthalate/polybutylene terephthalate
POA	Percent open area
PPSN	Polypropylene spun-bond nonwoven
PSD	Pore-opening size distribution
PVA	Polyvinyl alcohol
PVdF	Polyvinylidene fluoride
RMSE	Root mean square errors
RSM	Response surface methodology
SE	Shielding effectiveness
SEM	Scanning electron microscopy
SF	Silk fibroin
StdFD	Standard deviation of fiber diameter
TEM	Transmission electron microscopy

Preface

Nanotechnology, refers to a field whose theme is the control of matter on an atomic and molecular scale. Generally nanotechnology deals with structures of the size 100 nanometers or smaller, and involves developing materials or devices within that size. Nanotechnology is extremely diverse, ranging from novel extensions of conventional device physics, to completely new approaches based upon molecular self-assembly, to developing new materials with dimensions on the nanoscale, even to speculation on whether we can directly control matter on the atomic scale. There has been much debate on the future of implications of nanotechnology. Nanotechnology has the potential to create many new materials and devices with wide-ranging applications, such as in medicine, electronics, and energy production. On the other hand, nanotechnology raises many of the same issues as with any introduction of new technology, including concerns about the toxicity and environmental impact of nanomaterials, and their potential effects on global economics, as well as speculation about various doomsday scenarios. These concerns have led to a debate among advocacy groups and governments on whether special regulation of nanotechnology is warranted.

Nanotechnology is now used in precision engineering, new materials development as well as in electronics; electromechanical systems as well as mainstream biomedical applications in areas such as gene therapy, drug delivery and novel drug discovery techniques.

Nanofibers are defined as fibers with diameters on the order of 100 nanometers. They can be produced by interfacial polymerization and electrospinning. Nanofibers are included in garments, insulation and in energy storage. They are also used in medical applications, which include drug and gene delivery, artificial blood vessels, artificial organs and medical facemasks. Electrospinning is the cheapest and the most straightforward way to produce nanomaterials. Electrospun nanofibres are very important for the scientific and economic revival of developing countries. It is now possible to produce a low-cost, high-value, high-strength fibre from a biodegradable and renewable waste product for easing environmental concerns. Electrospun nanofibres can be used in many applications. This book presents new research in this dynamic field and covers all aspects of electrospinning as used to produce nanofibres.

— A. K. Haghi

Chapter 1

Electrospun Nanofibers: An Introduction

INTRODUCTION

An emerging technology of manufacturing of thin natural fibers is based on the principle of electrospinning process. In conventional fiber spinning, the mechanical force is applied to the end of a jet. Whereas in the electrospinning process the electric body force act on element of charged fluid. Electrospinning has emerged as a specialized processing technique for the formation of sub-micron fibers (typically between 100 nm and 1 μm in diameter), with high specific surface areas. Due to their high specific surface area, high porosity, and small pore size, the unique fibers have been suggested for wide range of applications. Electrospinning of natural fibers offers unique capabilities for producing novel natural nanofibers and fabrics with controllable pore structure. Current research effort has focused in understanding the electrospinning of natural fibers in which the influence of different governing parameters are discussed.

A schematic diagram to interpret electrospinning of polymer nanofibers is shown in Fig. 1.1. There are basically three components to fulfill the process: a high voltage supplier, a capillary tube with a pipette or needle of small diameter, and a metal collecting screen.

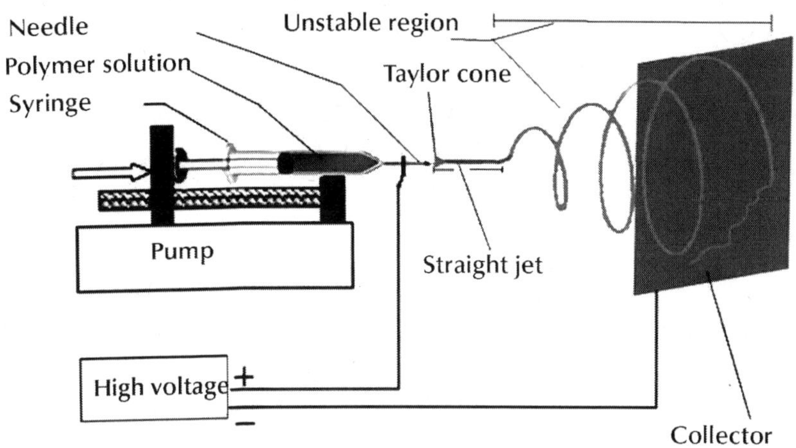

Figure 1.1. Electrospinning setup.

The morphological structure can be slightly changed by changing the solution flow rate as shown in Fig. 1.2. At the flow rate of 0.3 ml/h, a few of big beads were observed on fibers. The flow rate could affect electrospinning process. A shift in the mass-balance resulted in sustained but unstable jet and fibers with big beads were formed.

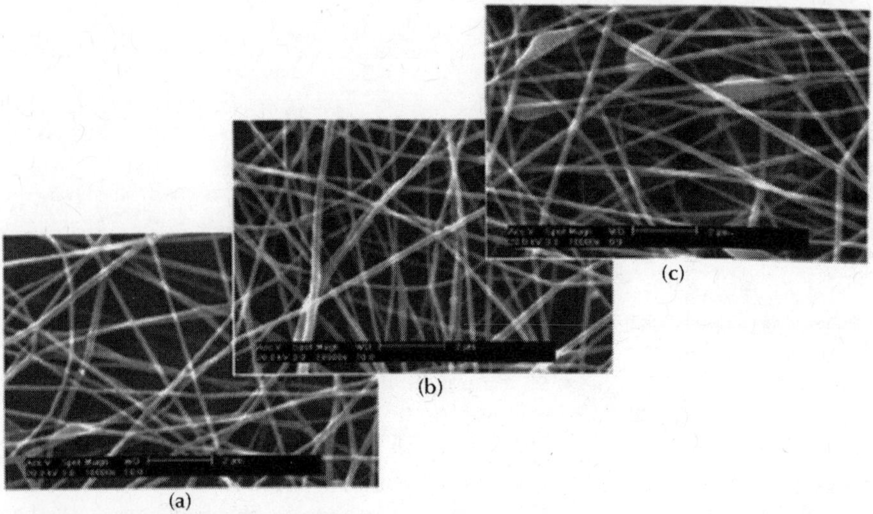

Figure 1.2. Effect of flow rate of 7% PVA water solution on fiber morphology (DH = 98%, voltage = 8 kV, tip–target distance = 15 cm). Flow rate: (a) 0.1 ml/h; (b) 0.2 ml/h; (c) 0.3 ml/h. Original magnification 10 k.

Typically, electrospinning has two stages. In the first, the polymer jet issues from a nozzle and thins steadily and smoothly downstream. In the second stage, the thin thread becomes unstable to a non-axisymmetric instability and spirals violently in large loops. The jet is governed by four steady-state equations representing the conservation of mass and electric charges, the linear momentum balance, and Coulomb's law for the E field. Mass conservation requires that

$$\pi R^2 v = Q \tag{1}$$

where Q is a constant volume flow rate. Charge conservation may be expressed by

$$\pi R^2 KE + 2\pi R v \sigma = I \tag{2}$$

where E is the z component of the electric field, K is the conductivity of the liquid, and I is the constant current in the jet. The momentum equation is formulated by Fig. 1.3:

$$\frac{d}{dz}(\pi R^2 \rho v^2) = \pi R^2 \rho g + \frac{d}{dz}[\pi R(-P + \tau_{ZZ})] + \frac{\gamma}{R}.2\pi R R' + 2\pi R(t_t^e - t_n^e R'), \tag{3}$$

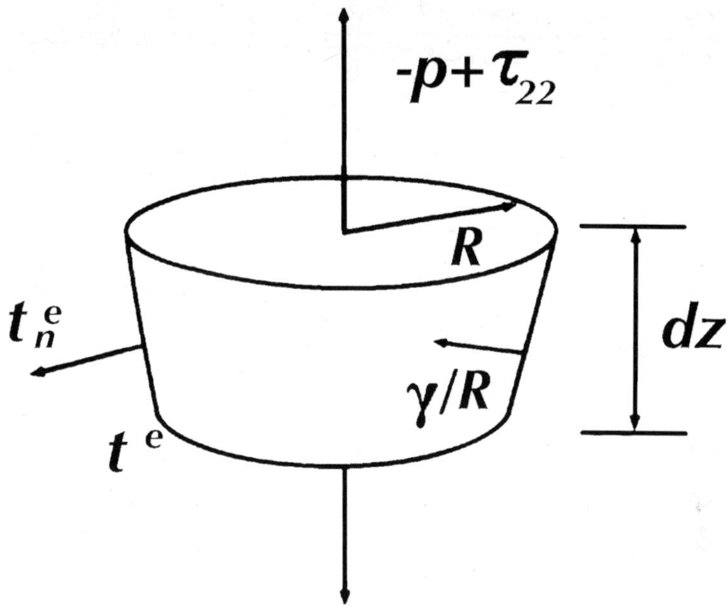

Figure 1.3. Momentum balance on a short section of the jet.

where τ_{zz} is the axial viscous normal stress, p is the pressure, γ is the surface tension, and t_t^e and t_n^e are the tangential and normal tractions on the surface of the jet due to electricity. The prime indicates derivative with respect to z, and R' is the slope of the jet surface. The ambient pressure has been set to zero. The electrostatic tractions are determined by the surface charge density and the electric field:

$$t_n^e = \left\| \frac{\varepsilon}{2}(E_n^2 - E_t^2) \right\| \approx \frac{\sigma^2}{2\overline{\varepsilon}} - \frac{\varepsilon' - \varepsilon}{2} E^2 , \tag{4}$$

$$t_t^e = \sigma E_t \approx \sigma E , \tag{5}$$

where ε and $\overline{\varepsilon}$ are dielectric constants of the jet and the ambient air, respectively, E_n and E_t are normal and tangential components of the electric field at the surface, and $\| * \|$ indicates the jump of a quantity across the surface of the jet. We have used the jump conditions for E_n and E_t: $\| \varepsilon E_n \| = \overline{\varepsilon E} - \varepsilon E_n = \sigma$, $\| E_t \| = \overline{E}_t - E_t = 0$, and assumed that $\varepsilon E_n << \overline{\varepsilon E}$ and $E_t \approx E$. The overbar indicates quantities in the surrounding air. The pressure $p(z)$ is determined by the radial momentum balance, and applying the normal force balance at the jet surface leads to:

$$-p+\tau_{rr}=t_n^e-\frac{\gamma}{R}, \quad (6)$$

Inserting Eqs. (4)–(6) into Eq. (3) yields:

$$\rho v v'=\rho g+\frac{3}{R^2}\frac{d}{dz}(\eta R^2 v')+\frac{\gamma R'}{R^2}+\frac{\sigma\sigma'}{\bar{\varepsilon}}+(\varepsilon-\bar{\varepsilon})EE'+\frac{2\sigma E}{R}, \quad (7)$$

Fig. 1.4 shows the relationship between mean fiber diameter and electric field with concentration of 15% at spinning distances of 5, 7, and 10 cm.

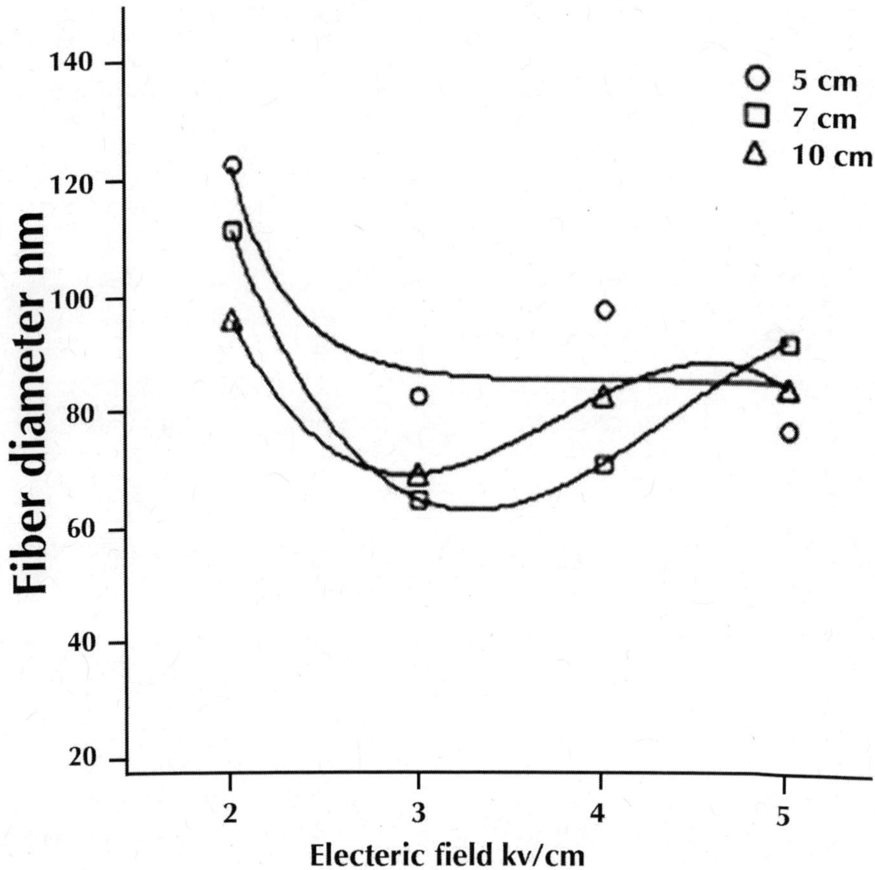

Figure 1.4. Momentum balance: The relationship between mean fiber diameter and electric field with concentration of 15% at spinning distances of 5, 7, and 10 cm.

The mean fiber diameter obtained at 2 kV/cm is larger than other electric fields.

KEYWORDS

- **Electrospinning**
- **Mass conservation**
- **Mean fiber diameter**
- **Momentum equation**

Chapter 2

Update on Effect of Systematic Parameters

INTRODUCTION

To date, a fundamental mechanism of the process of electrospinning is still characterized only qualitatively. The absence of a comprehensive theoretical knowledge of the electrospinning has resulted in polymer nanofibers with less controllable morphology and properties. A comprehensive study on this technique has been made in this chapter. Based on this study, many challenges exist in the electrospinning process of nanofibers, and a number of fundamental questions remain open.

Electrospun fibers are currently being utilized for several other applications as well. Some of these include areas in nanocomposites. Figure 2.1 compares the dimensions of nanofibers, micro fibers and ordinary fibers.

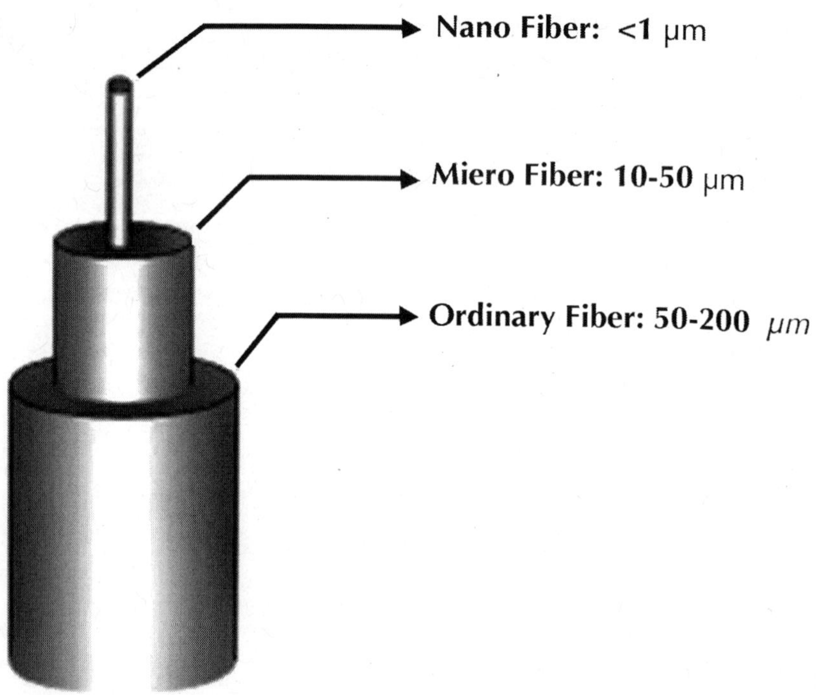

Figure 2.1. Classifications of fibers by the fiber diameter.

A schematic diagram to interpret electrospinning of nanofibers is shown in Fig. 2.2.

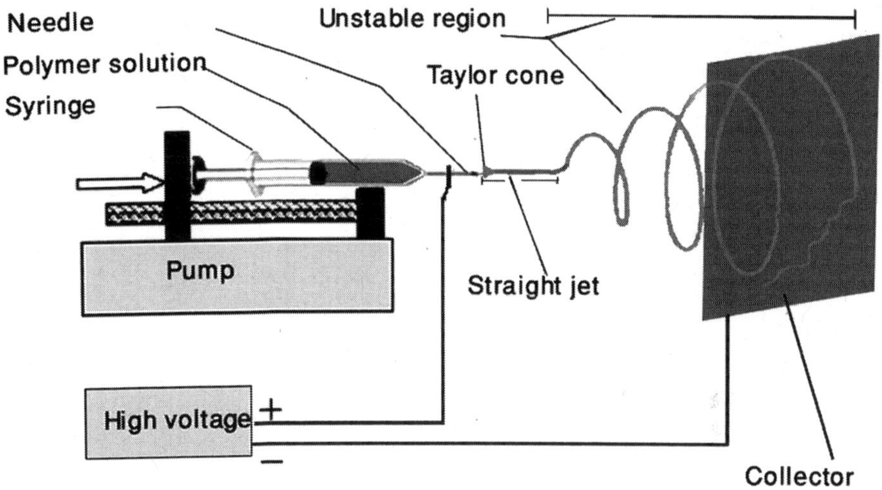

Figure 2.2. Electrospinning setup.

It has been found that morphology such as fiber diameter and its uniformity of the electrospun polymer fibers are dependent on many processing parameters. These parameters can be divided into three groups as shown in Table 2.1. Under certain condition, not only uniform fibers but also beads-like formed fibers can be produced by electrospinning.

Table 2.1. Processing parameters in electrospinning.

Solution properties	Viscosity Polymer concentration Molecular weight of polymer Electrical conductivity Elasticity Surface tension
Processing conditions	Applied voltage Distance from needle to collector Volume feed rate Needle diameter
Ambient conditions	Temperature Humidity Atmospheric pressure

EFFECT OF SYSTEMATIC PARAMETERS ON ELECTROSPUN NANOFIBERS

It has been found that morphology such as fiber diameter and its uniformity of the electrospun nanofibers are dependent on many processing parameters. These parameters can be divided into three main groups: (1) solution properties, (2) processing conditions, and (3) ambient conditions. Each of the parameters has been found to affect the morphology of the electrospun fibers.

Solution Properties

Parameters such as viscosity of solution, solution concentration, molecular weight of solution, electrical conductivity, elasticity and surface tension, have important effect on morphology of nanofibers.

Viscosity

The viscosity range of different nanofiber solution which is spinnable is different. One of the most significant parameters influencing the fiber diameter is the solution viscosity. A higher viscosity results in a large fiber diameter. Figure 2.3 shows the representative images of beads formation in electrospun nanofibers. Beads and beaded fibers are less likely to be formed for the more viscous solutions. The diameter of the beads become bigger and the average distance between beads on the fibers longer as the viscosity increases.

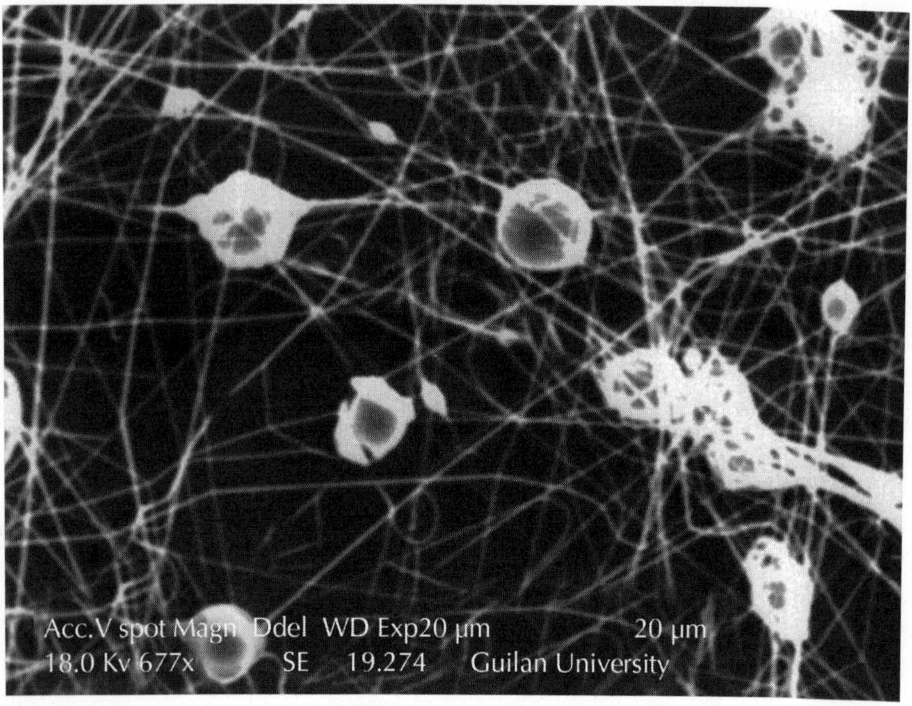

Figure 2.3. Electron micrograph of beads formation in electrospun nanofibers.

Solution Concentration

In electrospinning process, for fiber formation to occur, a minimum solution concentration is required. As the solution concentration increase, a mixture of beads and fibers is obtained (Fig. 2.4). The shape of the beads changes from spherical to spindle-like when the solution concentration varies from low to high levels.

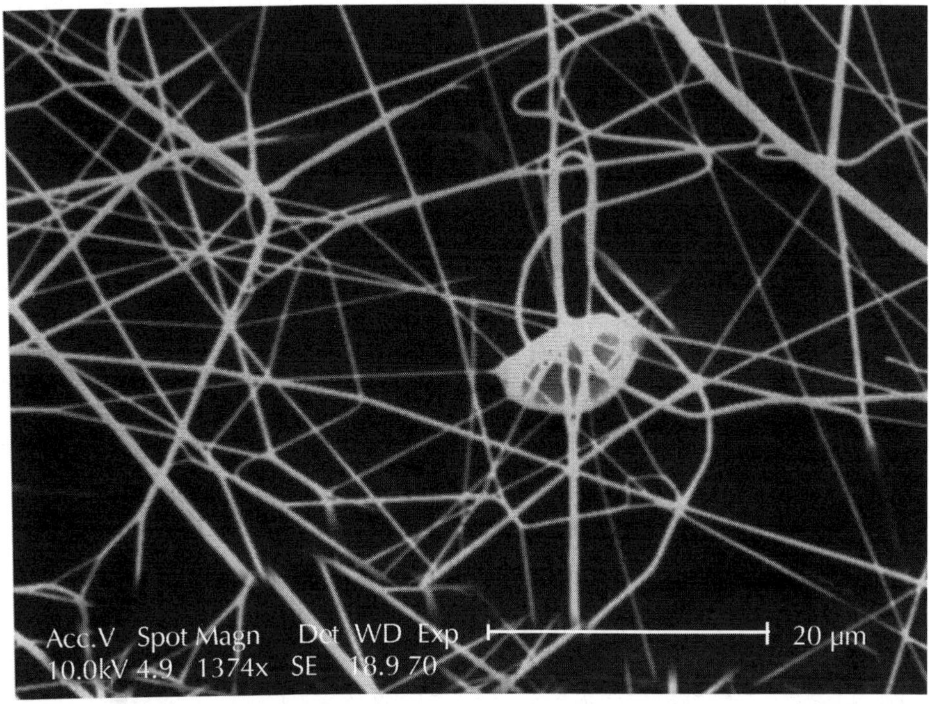

Figure 2.4. Electron micrograph of beads and fibers formation in electrospun nanofibers.

Molecular Weight

Molecular weight also has a significant effect on rheological and electrical properties such as viscosity, surface tension, conductivity and dielectric strength. It has been reported that too low molecular weight solution tend to form beads rather than fibers and high molecular weight nanofiber solution give fibers with larger average diameter (Fig. 2.5).

Surface Tension

By reducing surface tension of a nanofiber solution, fibers could be obtained without beads (Figs. 2.6 and 2.7). This might be correct in some sense, but should be applied with caution. The surface tension seems more likely to be a function of solvent compositions, but is negligibly dependent on the solution concentration. Different solvents may contribute different surface tensions. Droplets, bead and fibers can be driven by the surface tension of solution and lower surface tension of the spinning solution helps electrospinning to occur at lower electric field.

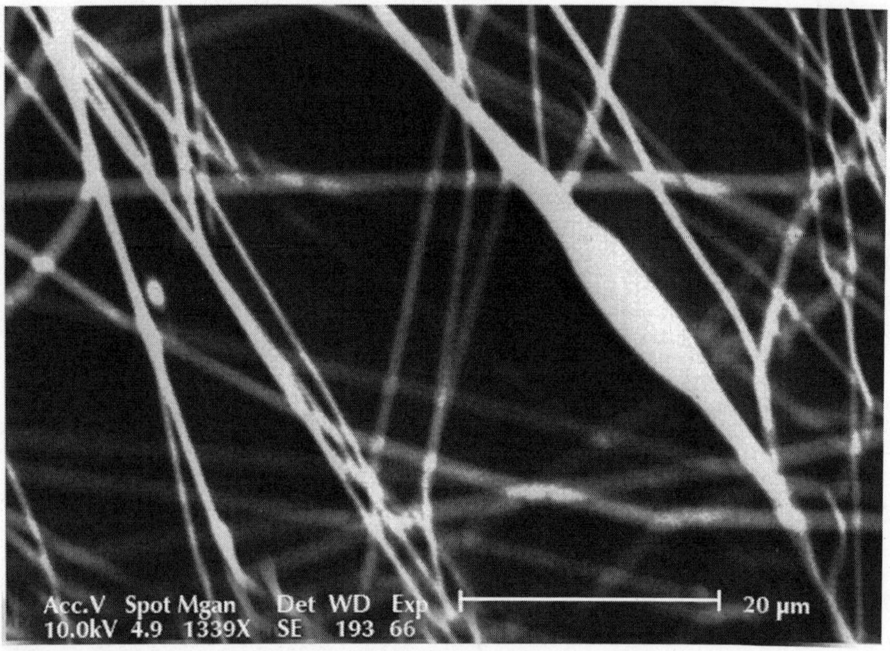

Figure 2.5. Electron micrograph of variable diameter formation in electrospun nanofibers.

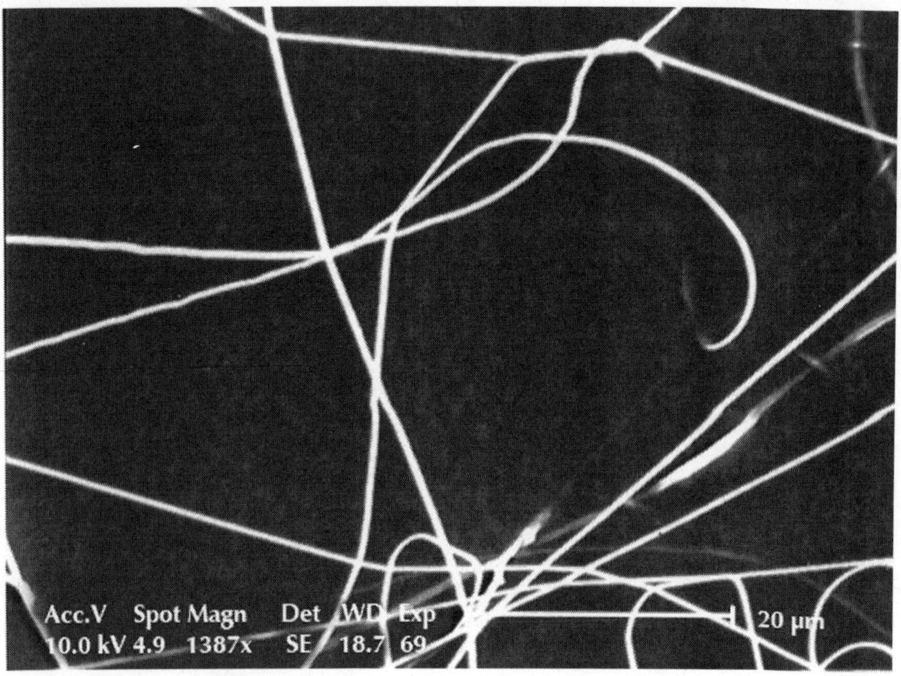

Figure 2.6. Electron micrograph of electrospun nanofiber without beads formation.

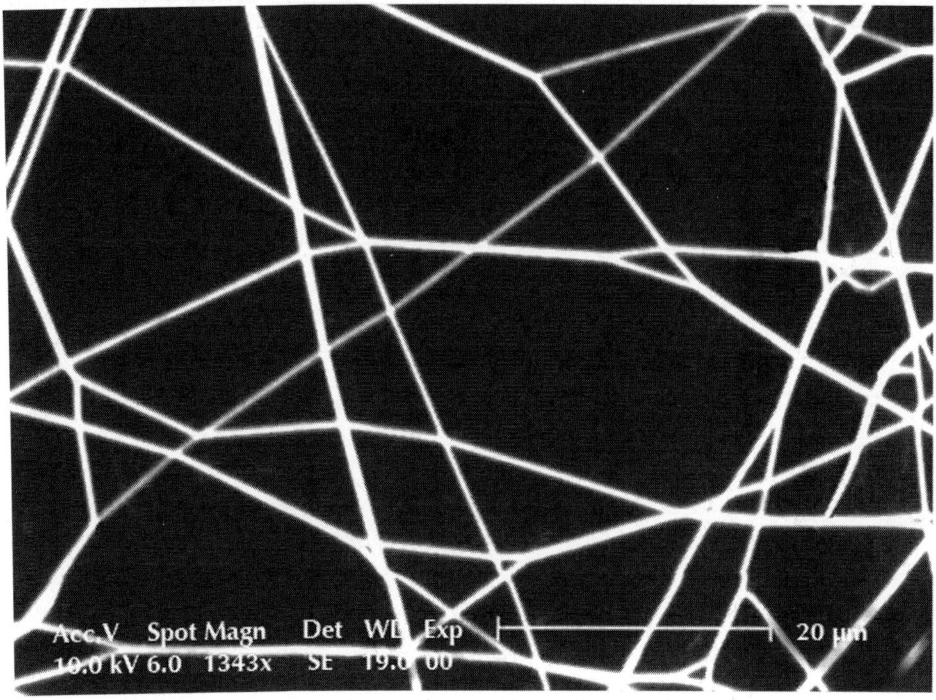

Figure 2.7. Electron micrograph of electrospun nanofiber without beads formation.

Solution Conductivity

There is a significant drop in the diameter of the electrospun nanofibers when the electrical conductivity of the solution increases. Beads may also be observed due to low conductivity of the solution, which results in insufficient elongation of a jet by electrical force to produce uniform fiber.

KEYWORDS

- **Dimensions of nanofibers**
- **Electrospun fibers**
- **Micro fibers**
- **Ordinary fibers**

Chapter 3

Update on Electrospun Polyacrylonitrile Nanofibers

INTRODUCTION

Polyacrylonitrile (PAN) nanofibers were prepared by electrospinning of PAN/DMF solutions of variable concentrations, applied voltages, feed rates and tip-target distances. The effects of all the above mentioned parameters on the electrospun PAN nanofibers are discussed in this chapter.

Our electrospinning apparatus consisted of a syringe pump, a 0–50 kV Dc power supply, an ammeter and various take up devices including metal screens, belts and other targets in an enclosed Faraday cage.

PAN fibers (thickness = 16 dtex; length = 150 mm ; bright) were used to prepare solutions. This polymer was dissolved in dimethylformamide (DMF) solvent at 40–50°C and different concentrations.

In concentration of 8 wt% (using 10 kv and 12 kv) formation of beaded fibers observed. Using 14kv and more, nanofibers without any beads were formed (Fig. 3.1).

In Fig. 3.1(c) the "fracture points" (indicated by arrows) are due to the jet resistance to any deformation during the whipping process.

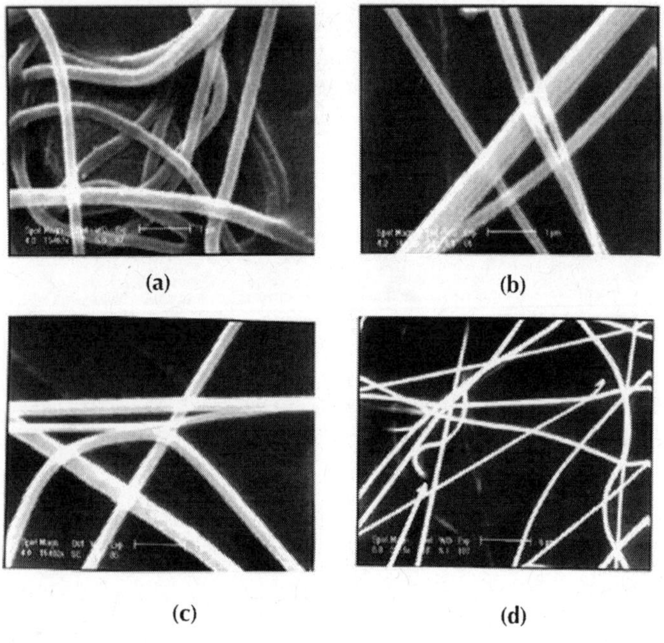

(a) (b) (c) (d)

Figure 3.1. SEM images of nanofibers produced from 8 wt% PAN/DMF, tip-targetdistance: 10 cm, feed rate: 0.5 ml/h, voltages used: (a) 10 kv, (b) 12 kv, (c) 14 kv, (d) 16 kv.

In this concentration (15 wt%), with increasing the applied voltage from 10 to 16 kv and holding the tip-target distance at 10 cm, the nanofibers exhibited that they could not completely dry. At 10 and 14 kv PAN webs with almost uniform diameter were obtained, but broader distributions in the diameter of nanofibers were resulted at 16 kv. Fracture points were obtained at 16 kv. No bead defect was present under these conditions (Fig. 3.2).

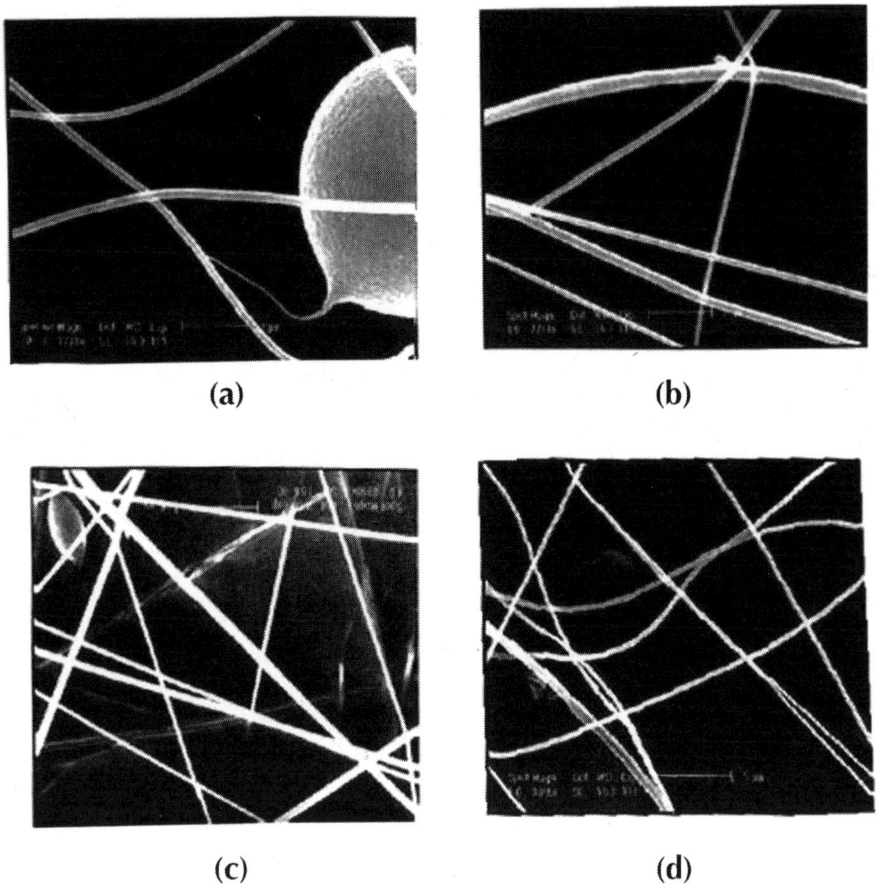

Figure 3.2. SEM images of nanofibers produced from 15 wt% PAN/DMF, tip-target distance: 15 cm, feed rate: 0.5 ml/h, voltage: (a) 10 kv, (b) 12 kv, (c)14 kv, (d) 16 kv.

In concentration of 15 wt%, at tip-target distance of 15 cm in comparison with 10 cm, the time for the solvent evaporate increased, and as a result, dry solid fibers were collected at the target. Distribution of diameter was higher at 12 kv and above, fracture points and beaded fibers were formed at 14 kv (Fig. 3.3).

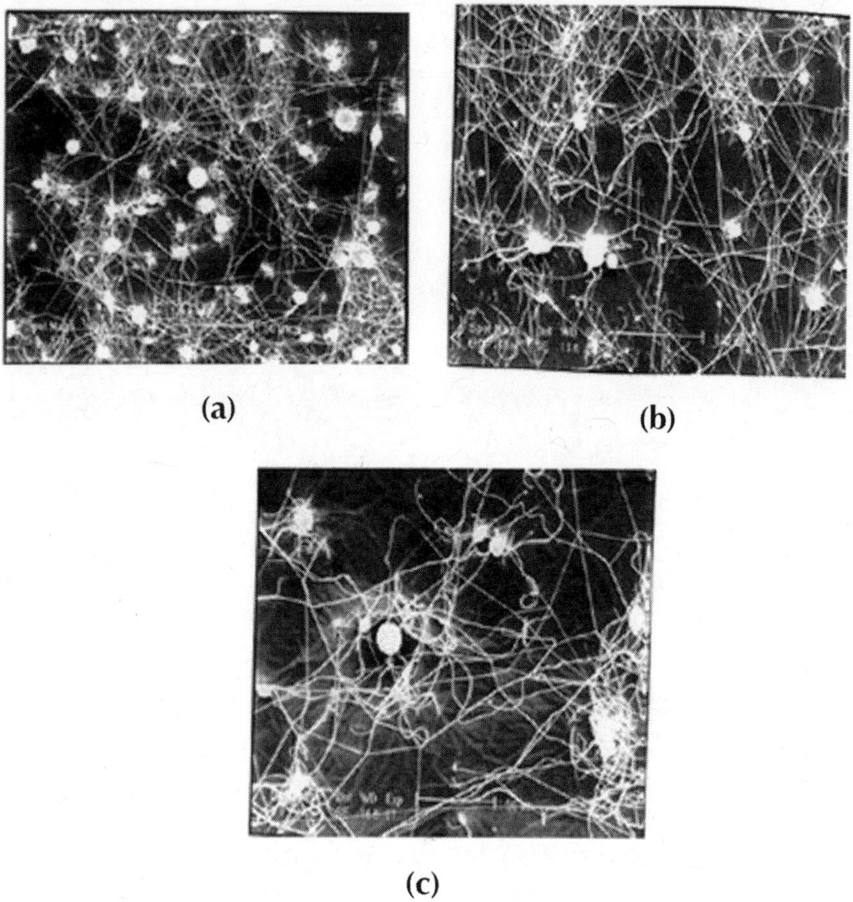

Figure 3.3. SEM images of beaded nanofibers produced from (a) 8 wt%, (b) 10 wt%,(c) 15 wt%, PAN/DMF. Tip-target distance: 10 cm, feed rate: 0.5 ml/h, voltage: 14 kv.

Drastic morphological changes were found when the concentration of the polymer solution was changed. In other words, the concentration or the corresponding viscosity was one of the most effective variables to control the fiber morphology. To investigate the effect of concentration on the fibers diameter, a series of experiments were carried out in which the concentration 8, 10, 15 wt% of PAN/DMF, with constant feeding rate 0.5 ml/h and tip-target distance of 10 cm. The applied voltages were varied from 10 to 16 kv.

CONCLUDING REMARKS

1. In 8 wt% at 14 kv and above fibers without any beads were formed and structure of nanofibers were not completely dry.
2. In 10 wt% uniform web without any beads were formed.

3. In 15 wt% at 15 cm tip-target distance in comparison with 10 cm the time for the solvent evaporate increased, with increased distance between the needle to collector.
4. Higher voltage needed to provide strong electric field and formation of thinner fibers.
5. It was observed that the diameter of the electrospun fibers was not dramatically changed with varied applied voltage.
6. With increasing concentration, the fibers diameter was observed to increase in all voltages except in 12 kv due to increasing the electrostatic forces which was not similar to rest.

KEYWORDS

- **Drastic morphological changes**
- **Electrospinning apparatus**
- **Fracture points**
- **Polyacrylonitrile (PAN) nanofibers**

Chapter 4

Update on Fabrication of New Class of Electrospun Nanofibers

INTRODUCTION

The use of fine fiber has become an important design tool for filter media. Nanofibers based filter media have some advantages such as lower energy consumption, longer filter life, high filtration capacity, easier maintenance, lower weight compared to other filter medias. The nanofibers based filter media made up of fibers with diameters ranging from 100 to 1000 nm conveniently produced by electrospinning technique. Typically, filter media are produced with a layer of fine fibers that can be used lonely or as a component in a media structure. The fine fiber increases the efficiency of filtration by trapping small particles which increases the overall particulate filtration efficiency of the structure. Improved fine fiber structures have been developed in this study in which controlled amount of fine fiber is placed on both sides of the media to result in an improvement in filter efficiency and substantial improvement in lifetime. In this research, regenerated silk fibroin obtained from industrial silk wastes was used to produce filter. Characteristics such as fibers diameter and its distribution, porosity and matt thickness of nanofiber filters which obtained in lab were examined by scanning electron microscopy (SEM) and using analyzed image processing algorithms.

Figures 4.1 and 4.2 illustrate the application of Fourier transform on a sample image with known orientation angles, 0°, 20°, and 90°. For the porosity analysis, SEM micrographs converted to binary format and then used (The picture pixels have only two values, 0 and 255).

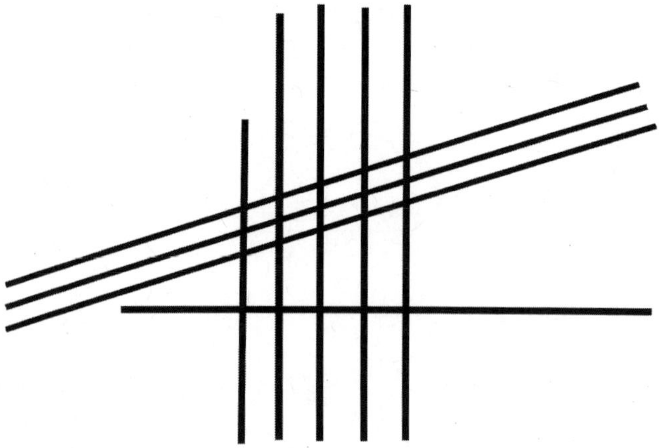

Figure 4.1. Sample image with known orientation angles, 0°, 20°, and 90°.

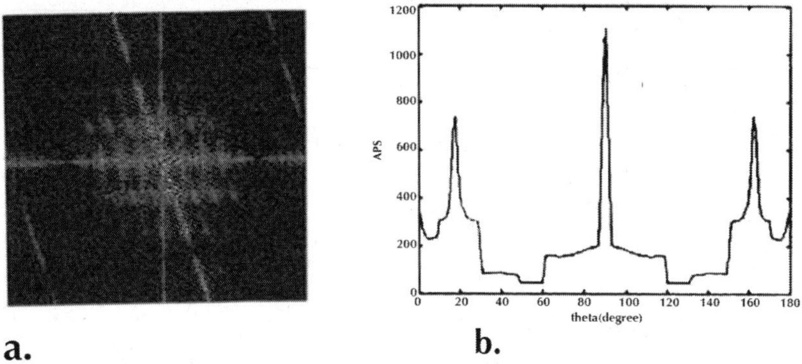

Figure 4.2. Fourier ransform and angular power spectrum of sample figure.

DIAMETER DISTRIBUTION OF NANOFIBERS

Diameter distribution of nanofibers and its average was extracted by using of Image program. Fig. 4.3 shows nanofibrous Medias obtained from solutions of 8 and 12 silk/ (formic acid) at the concentration of 12 wt%, the average fiber diameter is much larger than that of fibers spun at 8% concentration. The distribution of fiber diameters at 8 and 12 wt% concentrations is shown in Fig. 4.3. The fiber distribution becomes broader with increasing of concentration. Figure 4.4 shows same results for polyacrylonitrile (PAN) nanomats.

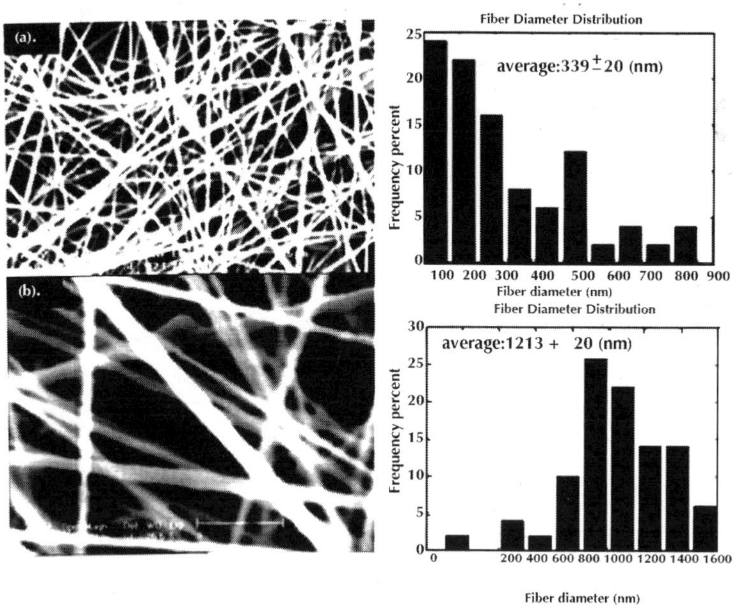

Figure 4.3. The distribution of fiber diameters and morphology of silk nanofibers at concentrations (a) 8 wt% and (b) 12 wt% constant tip-to collector distance of 7 cm and applied voltage 15 kV. Collector speed 100 r.p.m.

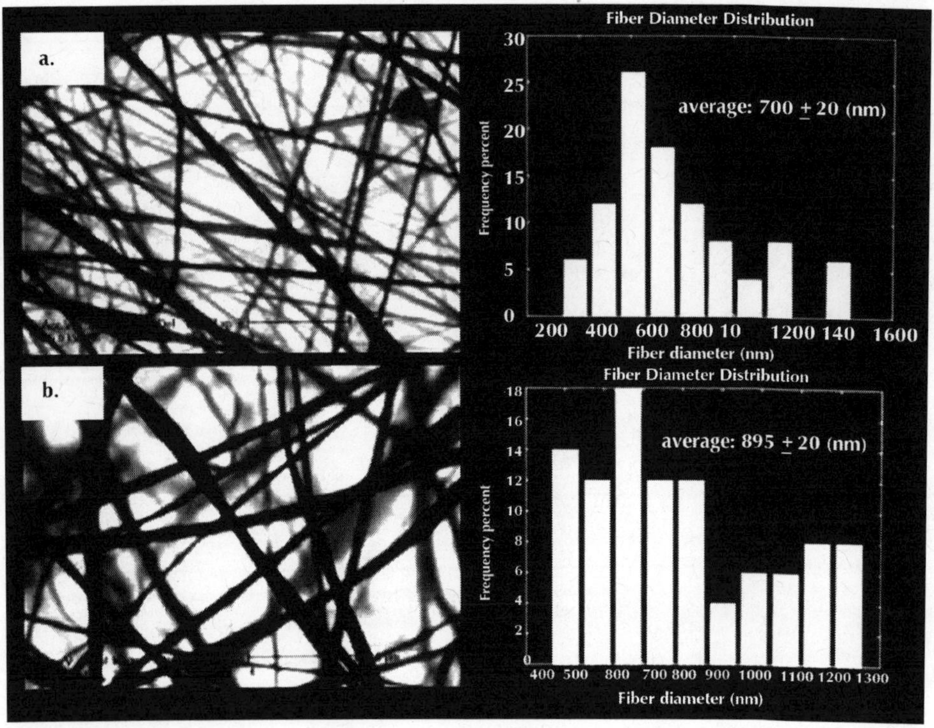

Figure 4.4. The distribution of fiber diameters and morphology of PAN nanofibers at concentrations (a) 8 wt% and (b) 13 wt% constant tip-to collector distance of 10 cm and applied voltage 12 kV. Collector speed 100 r.p.m.

NANOFIBER ORIENTATION DISTRIBUTION

The results for the Fourier transform and the orientation distribution of nanofibers (angular power spectrum histogram). As shown in Fig. 4.5, Fourier transform can detect the angular orientation (depicts orientation as a pick) of fibers with approximation.

For all samples of nanofiber-based nanomats produced at low concentration solutions were more uniform and arbitrary rather than sample from high concentration solutions.

CONCLUDING REMARKS

The porosity of nanofilters and the nanofiber diameter and its statistical parameters (average and distribution) were computed by analyzing of SEM pictures. The results indicated that increasing solution concentration leads to larger fiber diameter and broader diameter distribution in both silk and PAN nanofibers. Image analysis of porosity illustrated that in nanofiberious media with larger fiber diameter the porosity and empty spaces are much more than nanomats with finer nanofibers. It is clear that Fourier methods can provide good approximated value for the orientation distribution function (ODF) and can be a useful tool to characterization of nanofiberious media.

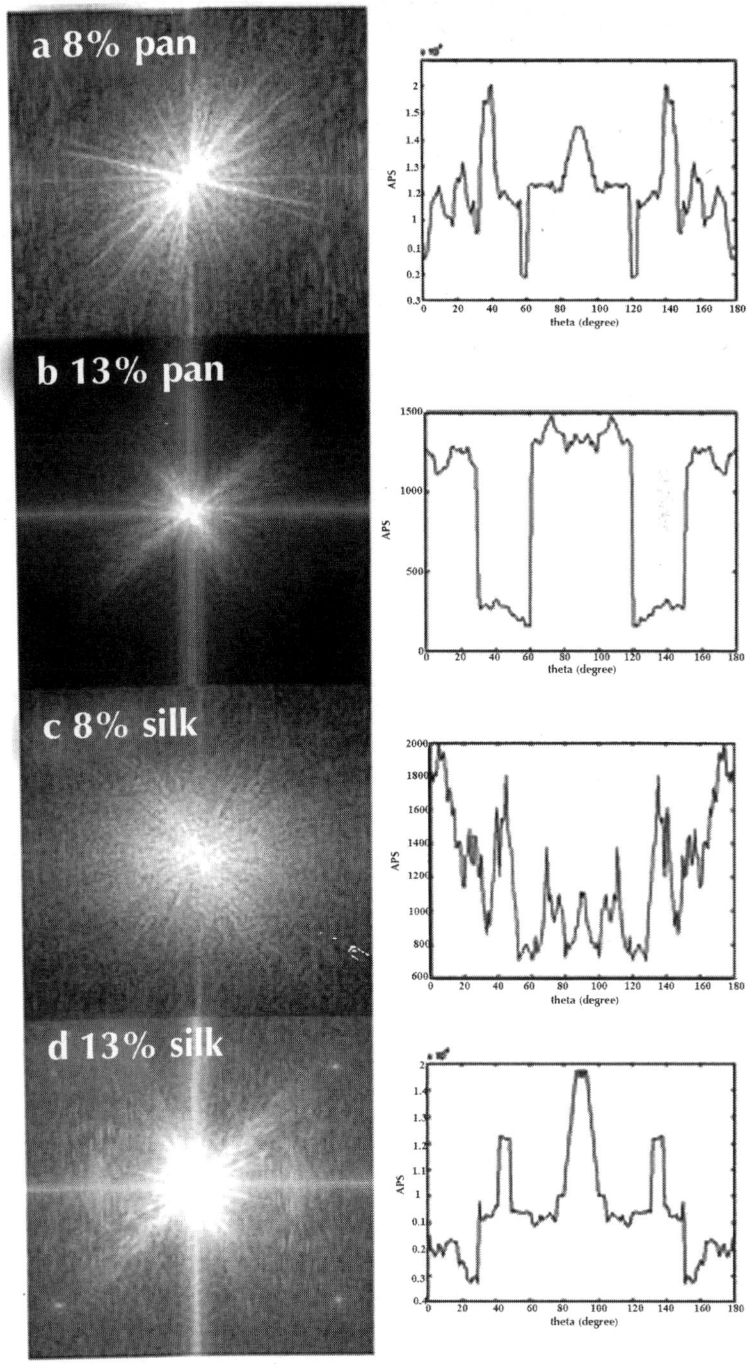

Figure 4.5. FFT transform and angular power spectrum of (a) PAN 8% concentration (fig. 5a), (b) PAN 13% concentration (fig. 5b) (c) silk 8% concentration (fig. 5c), (d) silk 13% concentration (fig.5d).

KEYWORDS

- **Diameter distribution of nanofibers**
- **Filter media**
- **Scanning electron microscopy**

Chapter 5

Update on Instability in Electrospun Nanofibers

INTRODUCTION

Electrospinning is a unique process to produce submicron polymeric fibers in the average diameter range of 50 to 500 nm. Polymer nanofibers can be made from variety of polymer solutions and are of substantial scientific interest including composite, filtration, protective clothing, biomedical, electronic applications, design of solar sails, and mirrors for use in space.

SEM examination of nanofibers collected on the aluminum plate revealed necking patterns and beads at certain places along the nanofibers. The uneven diameter and charge distribution along the electrospun jet is assumed to be the trigger of the growing perturbations leading to the formation of necks on electrospun nanofibers (Fig. 5.1). Due to the instability caused by high evaporation rate of the solvent, the fiber formed had larger diameter and beaded structural defects (Figs. 5.2–5.4). Correct selection of applied voltage, needle tip-collector distance, solution concentration, conductivity, and solvent volatility greatly influences the fiber structure and diameter (Figs. 5.5–5.6).

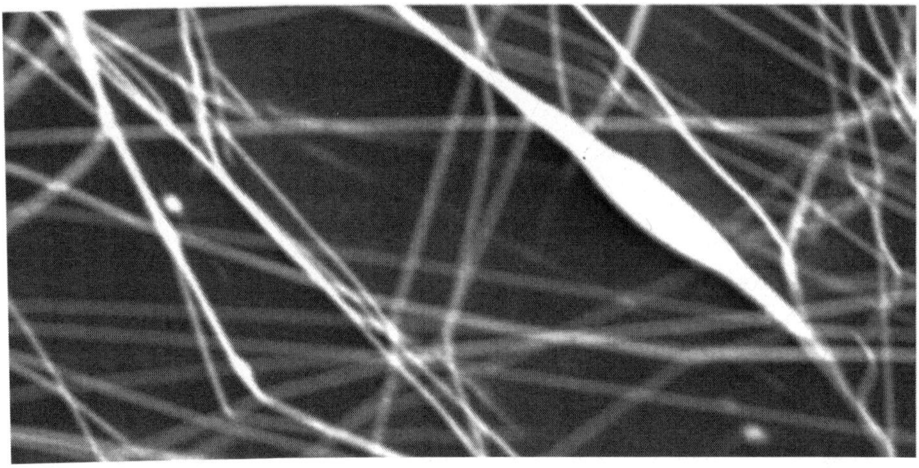

Figure 5.1. Formation of necks on nanofibers.

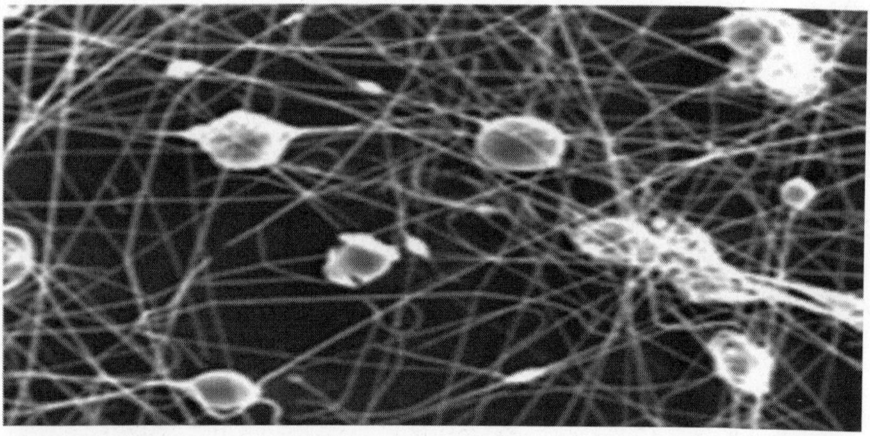

Figure 5.2. Formation of beads and defects due to instability.

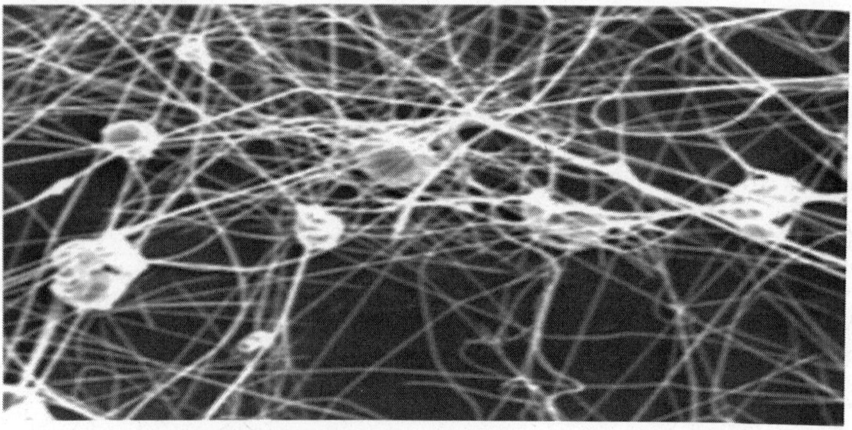

Figure 5.3. Formation of beads and defects due to instability.

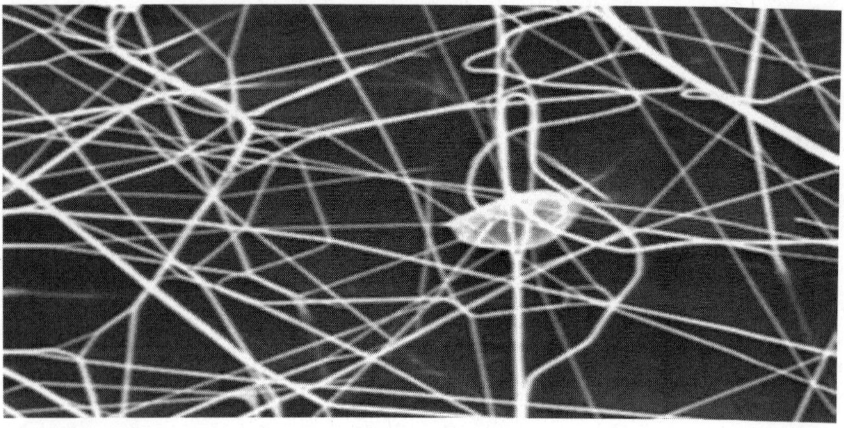

Figure 5.4. Formation of beads and defects due to instability.

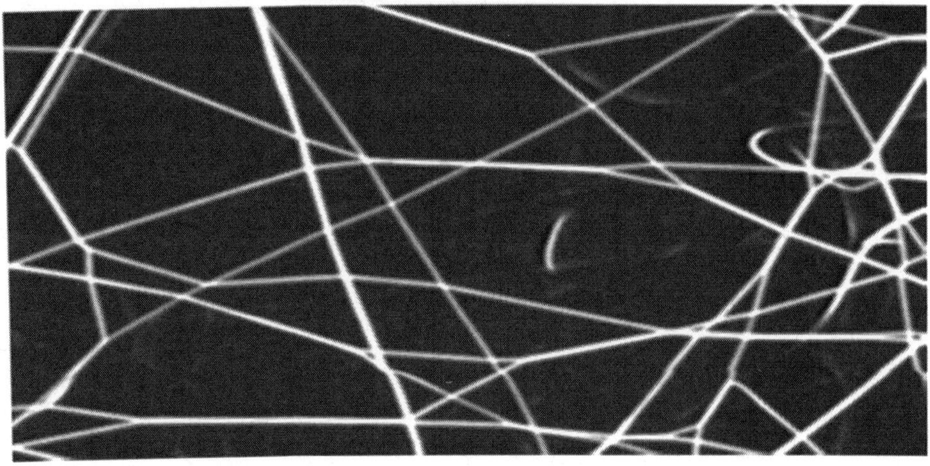

Figure 5.5. SEM images of nanofibers using correct setups.

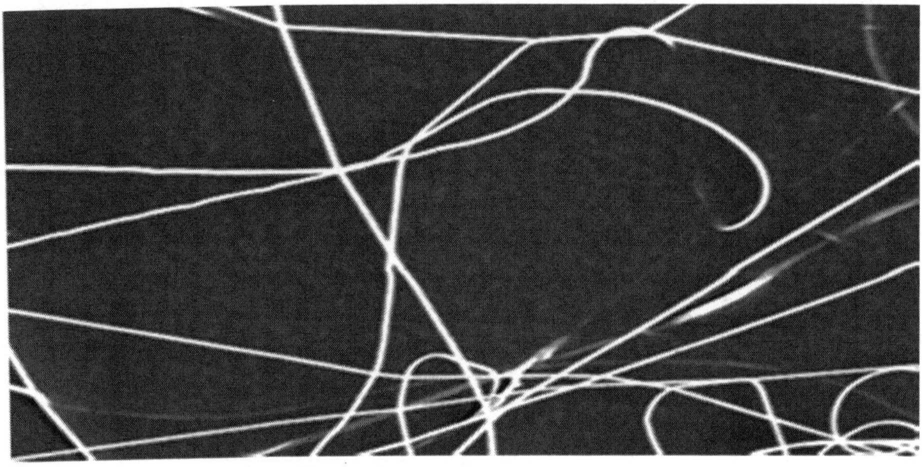

Figure 5.6. SEM images of nanofibers using correct setups.

CONCLUDING REMARKS

There are three categories of variables that influence the electrospun fiber diameter, including (1) polymer solution variables, (2) process variables, and (3) environmental variables. Examples of solution variables are viscosity or polymer concentration, solvent volatility, conductivity, and surface tension. Process variables consist of electric field strength, fluid flow rate, and distance between electrodes. Low molecular weight fluids form beads or droplets in the presences of an electric field, while high molecular weight fluids generate fibers.

KEYWORDS

- **Electrospun fiber diameter**
- **Polymer nanofibers**
- **Scanning Electron Microscopy**

Chapter 6

Update on Control of Electrospun Nanofiber Diameter—Part I

INTRODUCTION

The objective of this chapter is to use image analysis for measuring electrospun fiber diameter. Two methods are presented; *distance transform* and *direct tracking*. The methods are compared with conventional used manual method and tested with some samples with known characteristics generated by a simulation algorithm.

Figure 6.1 illustrates the electrospinning setup.

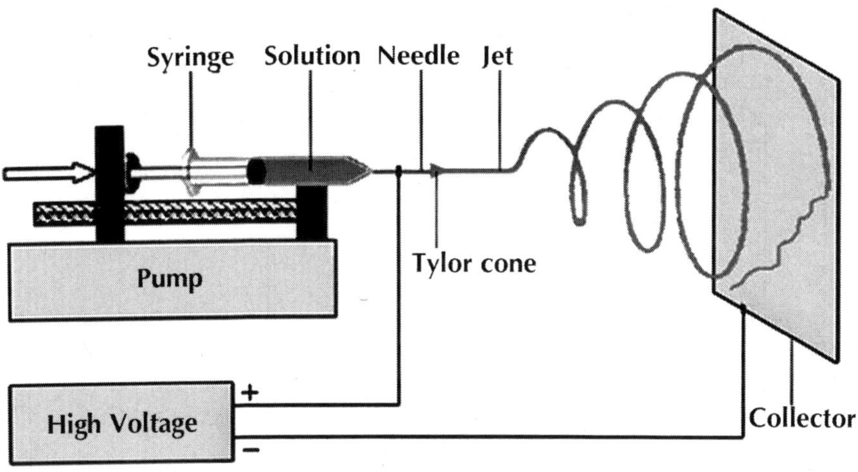

Figure 6.1. Electrospinning setup.

Simulation of Electrospun Web

The aim of the simulation is to obtain unbiased arrays which are spatially homogeneous. Lately, it was revealed that the best way to simulate nonwovens of continuous fibers is through the second method. For the continuous fibers, it is assumed that the lines are infinitely long so that in the image plane, all lines intersect the boundaries. Under this scheme (Fig. 6.2), a line with a specified thickness is defined by the perpendicular distance *d* from a fixed reference point O located in the center of the image and the angular position of the perpendicular α. Distance *d* is limited to the diagonal of the image.

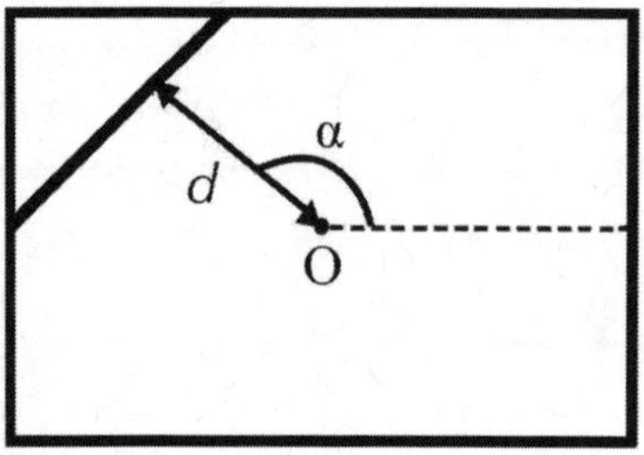

Figure 6.2. Procedure for μ-randomness.

Fiber Diameter Measurement

The first step in determining fiber diameter is to produce a high quality image of the web, called micrograph, at a suitable magnification using electron microscopy techniques. The methods for measuring electrospun fiber diameter are described in following sections.

MANUAL METHOD

The conventional method of measuring the fiber diameter of electrospun webs is to analyze the micrograph manually. The manual analysis usually consists determining the length of a pixel of the image (setting the scale), identifying the edges of the fibers in the image, and counting the number of pixels between two edges of the fiber (the measurements are made perpendicular to the direction of fiber-axis), converting the number of pixels to *nm* using the scale and recording the result. Typically 100 measurements are carried out (Fig. 6.3).

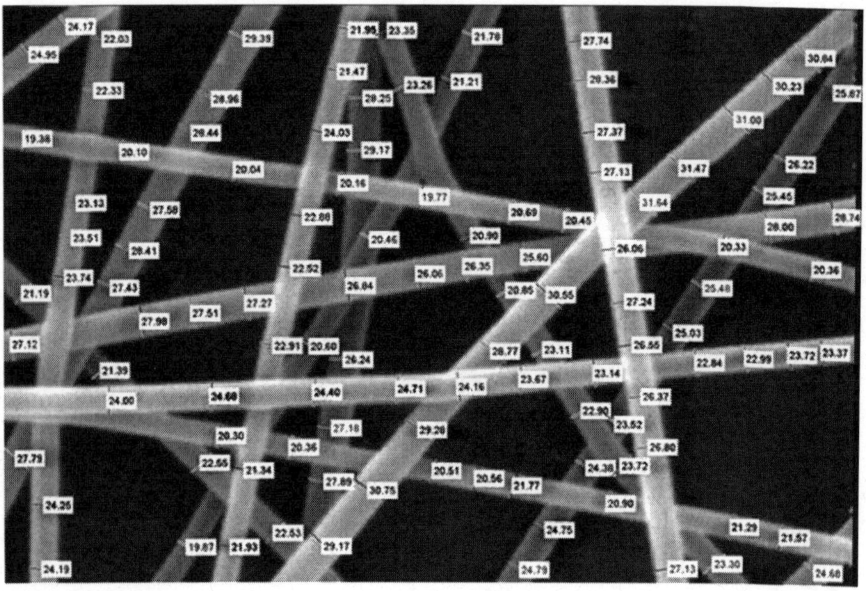

Figure 6.3. Manual method.

DISTANCE TRANSFORM

The *distance transform* of a binary image is the distance from every pixel to the nearest nonzero-valued pixel. The center of an object in the distance transformed image will have the highest value and lie exactly over the object's *skeleton*. The skeleton of the object can be obtained by the process of *skeletonization* or *thinning*. The algorithm removes pixels on the boundaries of objects but does not allow objects to break apart. This reduces a thick object to its corresponding object with one pixel width. Skeletonization or thinning often produces short spurs which can be cleaned up automatically with a *pruning* procedure (see Fig. 6.4).

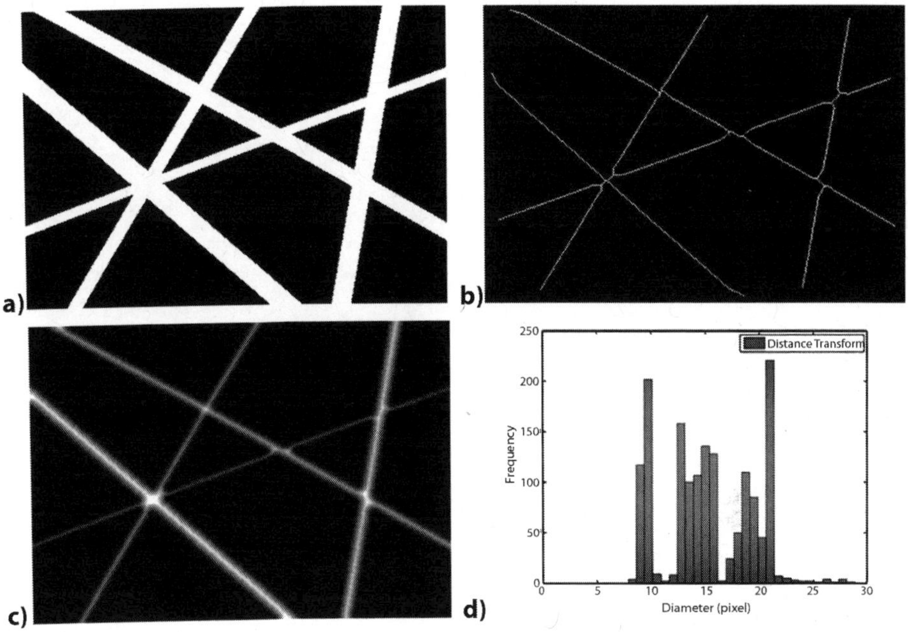

Figure 6.4. (a) A simple simulated image, (b) Skeleton of (a), (c) Distance map of (a) after pruning, (d) Histogram of fiber diameter distribution obtained by distance transform method.

DIRECT TRACKING

Direct tracking method uses a binary image as an input data to determines fiber diameter based on information acquired from two scans; first a horizontal and then a vertical scan. In the horizontal scan, the algorithm searches for the first white pixel adjacent to a black. Pixels are counted until reaching the first black. The second scan is then started from the mid point of horizontal scan and pixels are counted until the first black is encountered. Direction changes if the black pixel is not found. Having the number of horizontal and vertical scans, the number of pixels in perpendicular direction which is the fiber diameter could be measured from a geometrical relationship. The explained process is illustrated in Fig. 6.5.

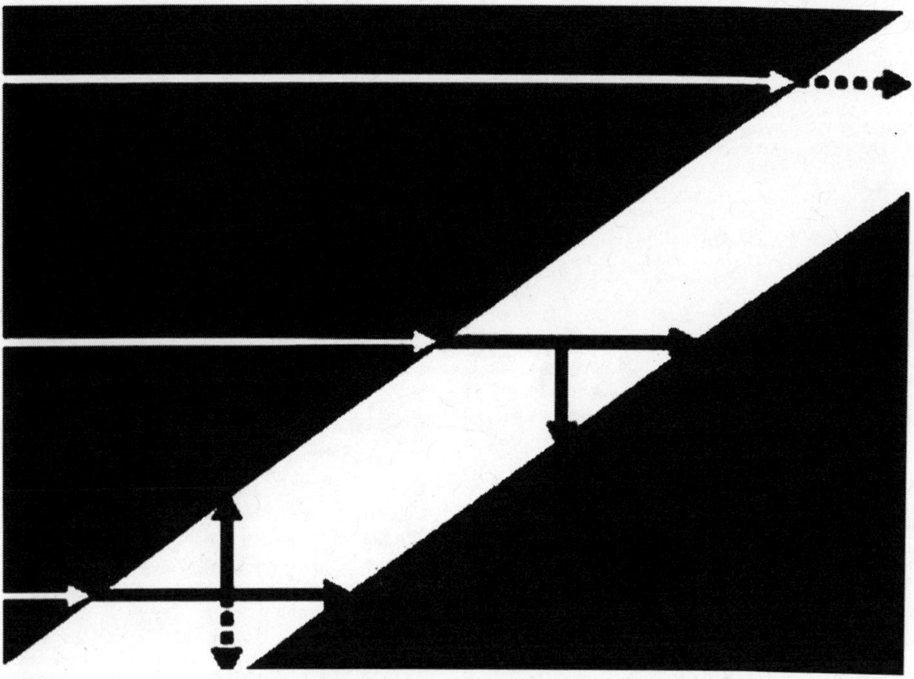

Figure 6.5. Diameter measurement based on two scans in direct tracking method.

Figure 6.6 shows a labeled simple simulated image and the histogram of fiber diameter obtained by this method.

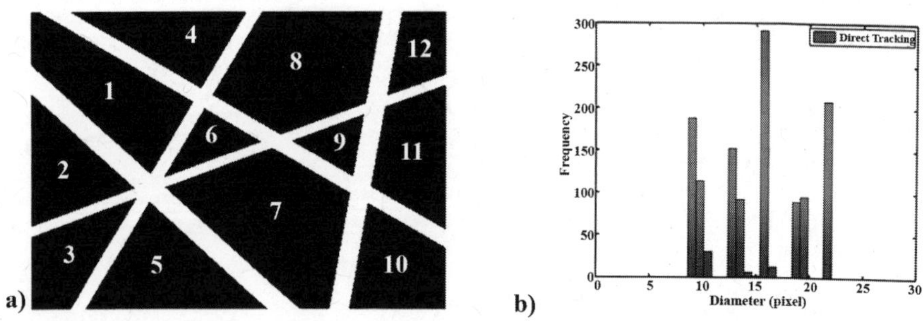

Figure 6.6. (a) A simple simulated image which is labeled, (b) Histogram of fiber diameter distribution obtained by direct tracking.

CONCLUDING REMARKS

Fiber diameter is the most important structural characteristics in electrospun nonwoven webs. The typical way of measuring electrospun fiber diameter is through manual method which is a tedious, time consuming and an operator-based method and cannot be used as an automated technique for quality control. The use of image analysis was investigated in this chapter for determining fiber diameter and developing of an automated method called direct tracking.

KEYWORDS

- **Direct tracking method**
- **Electrospun fiber diameter**
- **Skeletonization or thinning**

Chapter 7

Update on Control of Electrospun Nanofiber Diameter—Part II

INTRODUCTION

In this chapter, a new distance transform method for measuring fiber diameter in electrospun nanofiber webs has been described. In this algorithm, the effect of intersection has been eliminated which brings more accuracy to the measurement. The effectiveness of the method was evaluated by a series of simulated images with known characteristics, as well as some real webs obtained from electrospinning of PVA. The new method was then compared with the original distance transform method. The results obtained by the new method were significantly superior than the distance transform, indicating that the new method could successfully be used to measure electrospun fiber diameter. Figure 7.1 illustrates the electrospinning setup.

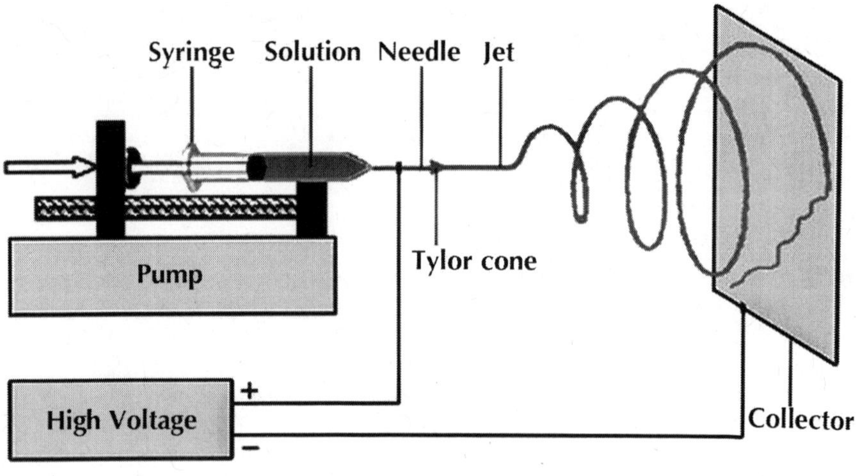

Figure 7.1. Electrospinning setup.

The properties of electrospun nanofiber webs depend not only on the nature of the component fibers but also on its structural characteristics. In the last few years, image analysis methods have been developed in order to identify fibers and measure nonwoven characteristics such as fiber orientation, fiber diameter, pore size, uniformity, and other structural features. Note however that, since these are new techniques and their accuracy and limitations have not been verified, samples with known characteristics are required to evaluate the accuracy of the methods which can be produced by simulation schemes.

Fiber diameter is the most important structural characteristic in electrospun nanofiber webs. Despite the importance, thus far there is no successful method for determining fiber diameter and a few works have been conducted to develop a method for measuring fiber diameter. Furthermore, large scale production of nanofibers requires unique online quality control. Hence, developing an accurate and automated fiber diameter measurement technique is useful and crucial. In a method image analysis has been used to measure fiber diameter in nonwoven textiles. Nevertheless, the method has some problems at the intersections of fibers making it inefficient for measuring electrospun nanofiber diameter. In this contribution, an attempt has been made to circumvent the problems associated with this method thereby developing a reliable, efficient, and automated method for measuring nanofiber diameter in electrospun webs.

METHODOLOGY

Fiber Diameter Measurement

Understanding how fiber diameter and its distribution are affected by the electrospinning variables is essential to produce nanofibers with desired properties. The extremely small fiber size and random production of nanofiber made its diameter measurement very difficult. Most commercially available measurement equipment cannot work with nanofibers. In order to measure fiber diameter, images of the webs are required. These images called micrographs usually are obtained by Scanning Electron Microscopy (SEM), Transmission Electron Microscopy (TEM), or Atomic Force Microscopy (AFM). Dealing with fiber diameter requires high-quality images with appropriate magnifications. The methods for measuring fiber diameter are presented as follows.

Manual Method: Routine measurement of fiber diameter and its distribution are carried out by manual method using micrographs obtained from SEM. First, the length of a pixel in the image is determined; in other words the scale is set. Then, a fiber is selected and pixels between two edges of the fiber perpendicular to the fiber axis are counted. The number of the pixels is then converted to *nm* using the scale and the resulting diameter is recorded. This procedure is repeated for other selections until any fiber is processed. Typically 100 diameters are measured (Fig. 7.2). Finally the histogram of fiber diameter distribution is plotted.

This process is very time-consuming and operator consistency and fatigue can reduce the accuracy. Identifying the edges of the fibers needs attention and the measurements are not exactly made perpendicular to the fiber axis. Furthermore, since it is an operator-based method, it cannot be used as online method for quality control. Automating the fiber diameter measurement which eliminates the use of an operator is a natural solution to this problem.

Distance Transform Method: The *skeleton* of an object in a binary image, which provides helpful information about the shape of the object, is defined as the corresponding object with one-pixel width. There are two approaches for assessing the object's skeleton: *skeletonization* and *thinning*. In the first, using medial axis transformation (MAT), the center points of the object which are equidistant from two closest points of the object's boundary are attained and set to as skeleton. Whereas, in the second, the pixels on the boundary of the object are removed without allowing it

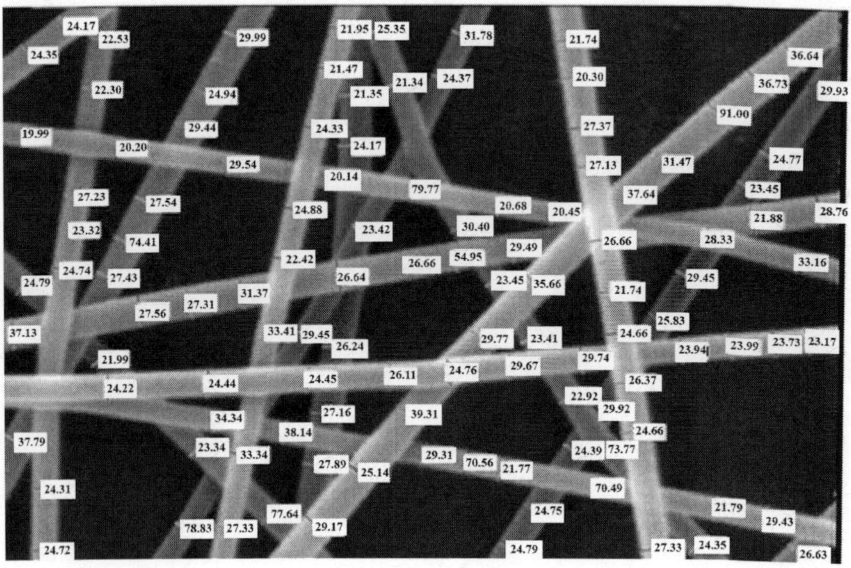

Figure 7.2. Manual method.

to break apart thereby shrinking a thick object to a centrally located one-pixel width object. In thinning operation, the following conditions must be satisfied:

1. An object must not break into pieces.
2. The end points must not be removed so that the object does not become shorter.
3. An object must not be deleted.

Both of these two operations result in line-like structures with one-pixel in thickness preserving the topology of the object. However, the skeleton obtained by skeletonization is often different with that of thinning and has more branches. Figure 7.3a shows a binary image to which skeletonization and thinning is applied and the resultant skeletons are depicted in Figs. 7.3b and 7.3c respectively. Note that these operations often produce short *spurs* (also called *parasitic components*) which may further be cleaned up by a postprocessing called *pruning* procedure (Fig. 7.3d). Identifying and removing the spurs iteratively, this procedure is an essential complement to skeletonization and thinning.

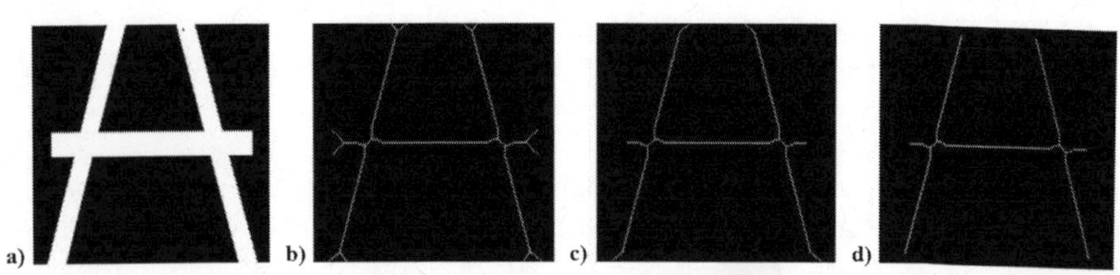

Figure 7.3. Obtaining the skeleton of a binary image: (a) a binary image, (b) skeletoninzation, (c) thinning, (d) resulting skeleton after pruning.

The distance transform is an operation which is applied to a binary image consisting of 1s and 0s corresponding to objects and background respectively, and results in a grayscale image often called *distance map* (known also as distance transformed image). For each pixel in the binary image, the corresponding pixel in the distance map has the value equal to the minimum distance between that pixel and the closest object pixel, that is to say, the distance from that pixel to the nearest non-zero valued pixel. There are several different sorts of distance transform according to which distance metric is being used in order to measure the distance between the pixels. Three common distance metric used in this approach are: city block, chessboard, and Euclidean. The city block distance gives the length of a path between the pixels according to a 4-connected neighborhood (moving only in horizontal and vertical directions). The city block distance between (x_1, y_1) and (x_2, y_2) is given by:

$$D_{\text{Cityblock}} = |x_1 - x_2| + |y_1 - y_2| \tag{1}$$

In contrast, the chessboard distance metric measures the path between the pixels based on an 8-connected neighborhood (diagonal move is also allowed) as if a King moves in chess. This metric is given by:

$$D_{\text{Chessboard}} = Max(|x_1 - x_2|, |y_1 - y_2|) \tag{2}$$

With the city block metric, distances in the direction of diagonals are longer resulting in diamond-shaped structures. In the case of chessboard metric is used. Square-shaped structures are obtained. Even though they could be used in certain applications, Euclidean metric is more practical and relevant, since it is the only one that preserves the isotropy of the continuous space (Fig. 7.4). The Euclidean distance which is the straight line distance between two pixels is defined as:

$$D_{\text{Euclidean}} = \sqrt{(x_1 - x_2)^2 + (y_1 - y_2)^2} \tag{3}$$

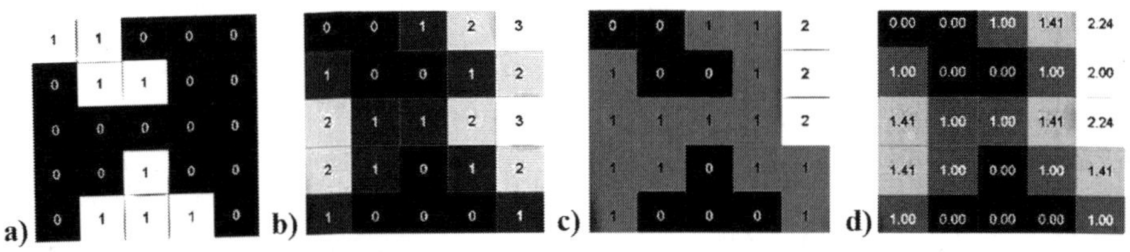

Figure 7.4. Distance map of a binary image: (a) a small binary image and its distance map obtained by, (b) city block, (c) chessboard, (d) Euclidean metric.

The center of an object in distance transformed image has the highest value which coincides with the axis of the object. Interestingly enough, an object's skeleton will lie exactly over the maximum of the distance transform for that object. This fact is clearly demonstrated in Fig. 7.5. Note that the z-position of the skeleton in Fig. 7.5d is quite arbitrary, just for showing the coincidence. Serving as the basic component of the methods, this remarkable feature will later be utilized for determining nanofiber diameter distribution in electrospun webs.

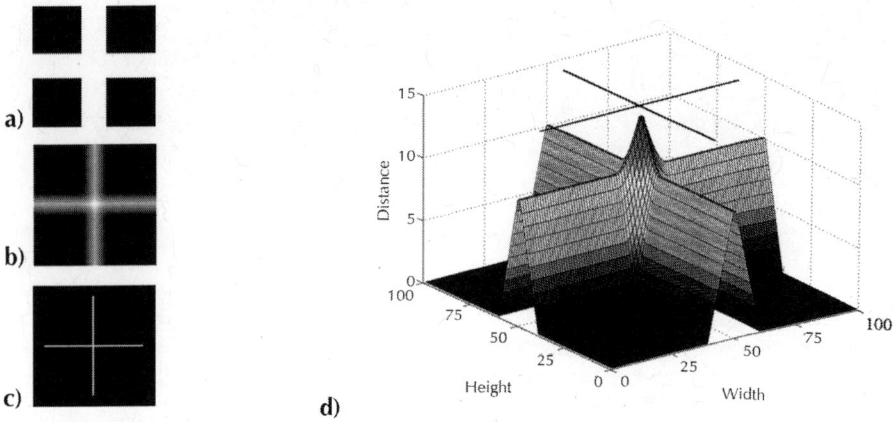

Figure 7.5. Skeleton lies exactly over the center of distance map: (a) a simple 100 x 100 binary image, (b) Euclidean distance map, (c) skeleton obtained by thinning after pruning, (d) 3-D plot showing the coincidence of skeleton and center of the object in distance map.

The algorithm for determining fiber diameter uses the skeleton and distance map of binary input image. Since our images consist of light fibers on dark background, they first need to be complemented. In the complement of a binary image, zeros become ones and ones become zeros; black and white are reversed. Thus fibers become black and background white. The complemented image is used to create a distance transformed image. Then the skeleton of the objects is created from the input binary image by the process of skeletonization or thinning. Fiber diameter is then determined using the distance transformed image and the skeleton. The skeleton acts as a guide for tracking the distance map and distances at all points along the skeleton (which coincide with the center of the objects in the distance transformed image) are recorded to compute fiber diameters. Finally, the recorded results are doubled and fiber diameters are obtained. The values (in pixels) may further be converted to *nm* and the histogram of fiber diameter distribution is plotted. Figure 7.6 shows a simple simulated image together with its skeleton and distance map including the histogram of fiber diameter obtained by this method.

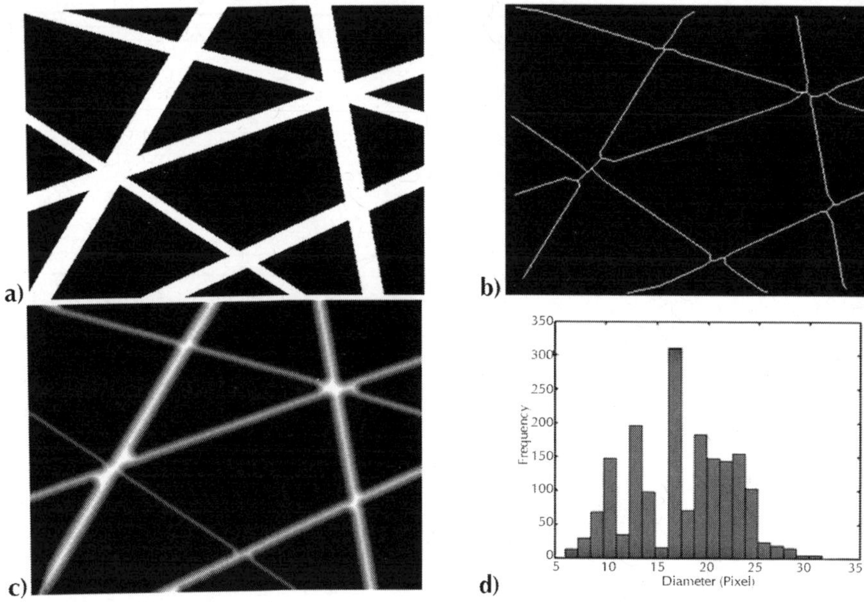

Figure 7.6. Distance transform method: (a) a simple simulated image, (b) skeleton of (a), c) distance map of (a) after pruning, (d) histogram of fiber diameter distribution.

New Distance Transform Method: As it is depicted in Fig. 7.7a, the intersections in the distance map are brighter than where a single fibers present. This demonstrates that higher values than expected were returned at these points. Figure 7.7b shows the broken skeleton at intersections. This problem becomes more pronounced as fibers get thicker and for points where more fibers cross each other. Hence, the distance transform method fails in measuring fiber diameter at intersections.

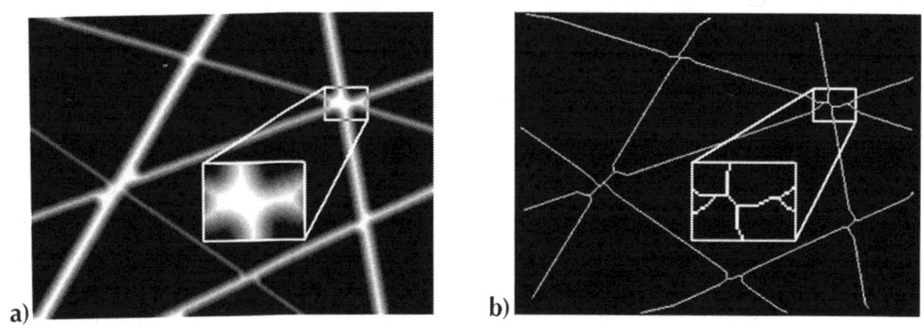

Figure 7.7. Distance transform method failure at intersection points: (a) distance map of the image shown in Fig. 7.6a, (b) broken skeleton obtained from thinning of Fig. 7.6a (area around an intersection has been magnified for more clarity).

We modified the distance transform method so that the problems associated with the intersections are solved. Furthermore city block distance transform was used which as mentioned earlier, is not a realistic metric, since it does not preserve the isotropy. In order to provide more rational results, in this approach we used Euclidean distance metric. The method uses a binary image as an input. Then, the distance transformed image and its skeleton are created. In order to solve the problem of the intersections, these points are identified and deleted from the skeleton.

First, in order to find the intersection points, a *sliding neighborhood operation* is employed. A sliding neighborhood operation is an operation which is applied to a pixel at a time; the value of that pixel in the output image is determined by the implementation of a given function to the values of the corresponding input pixel's neighborhood (Fig. 7.8). A neighborhood about a pixel, which is usually called the center point, is a square or rectangular region centered at that pixel. The operation consists of five steps:

1. Defining a center point and a neighborhood block.
2. Starting from the first (normally top left) pixel in the image.
3. Performing an operation (a function given) that involves only the pixels in the defined block.
4. Finding the pixel in the output corresponding to the center pixel in the block and setting the result of the operation as the response at that pixel.
5. Repeating steps 3 to 4 for each pixel in the input image.

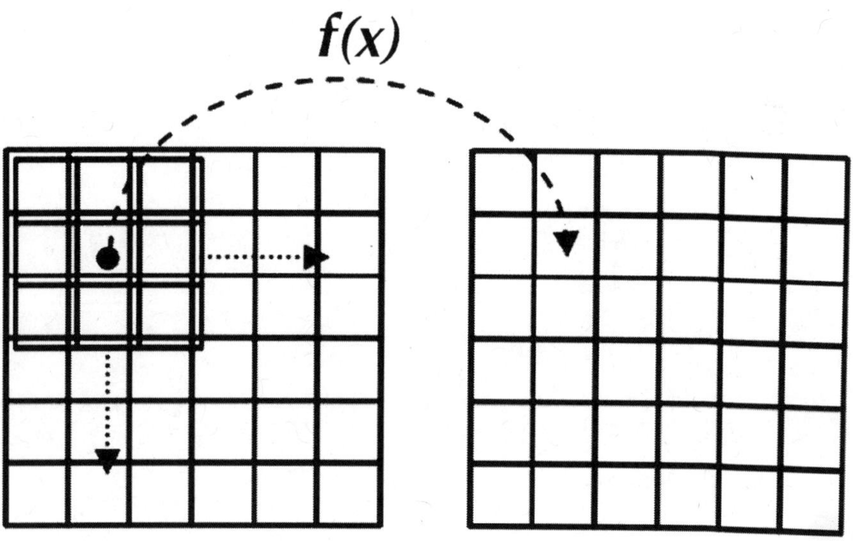

Figure 7.8. Sliding neighborhood operation with a 3-by-3 neighborhood block.

Since at an intersection point, two or more fibers meet each other, it could be defined as a location where a white pixel in the skeleton has more than two neighboring pixels each leading a branch. Hence, performing a sliding neighborhood operation on the skeleton with a 3-by-3 sliding block and summation as the function (which is applied over all pixels in the block), the intersections could be identified as the points having values more than 3. This is demonstrated in Fig. 7.9 (the intersections are shown with arrows).

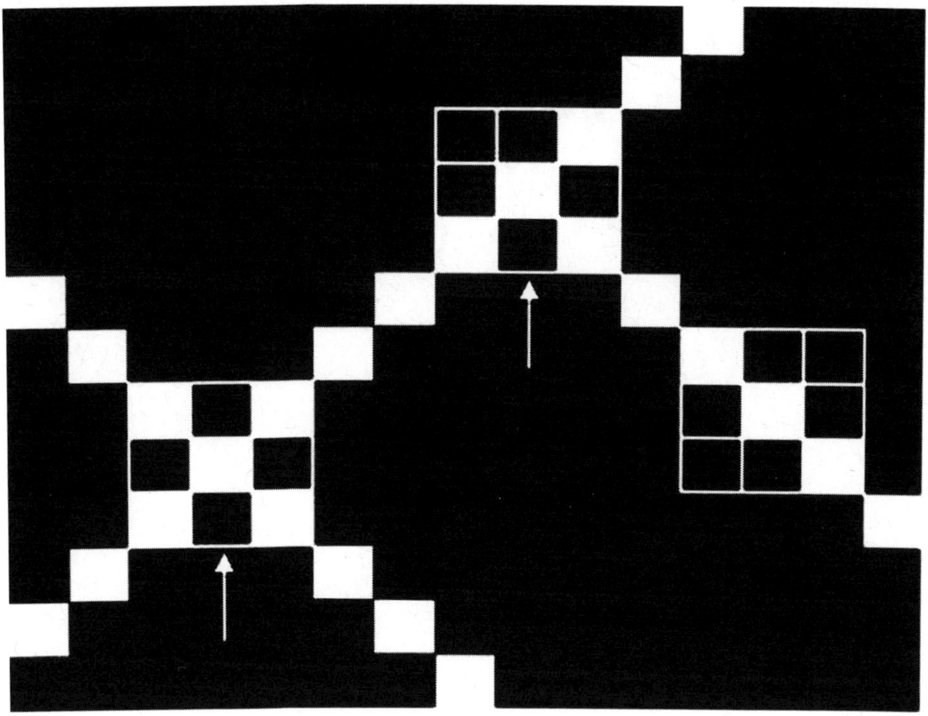

Figure 7.9. Identifying intersection points using a sliding neighborhood operation with a 3-by-3 neighborhood block.

After the intersection points are located, the next step is to find the width of each one. This is carried out using the distance map of the binary input image via finding the pixel corresponding to that intersection point.

Finally, the resultant skeleton (of which the intersections are deleted) is used as a guide for tracking the distance transformed image and fiber diameters are obtained by recording the intensities to at all points along the skeleton (white pixels in Fig. 7.10a show the skeleton) and doubling the results. The distance map of image in Fig. 7.6a is also shown in Fig. 7.10b for better understanding of the procedure. Setting the length of a pixel in the image, the values may then be converted to *nm* and the histogram of fiber diameter distribution is plotted. Figure 7.10c demonstrates the histogram of fiber diameter (in term of pixel) obtained by this method.

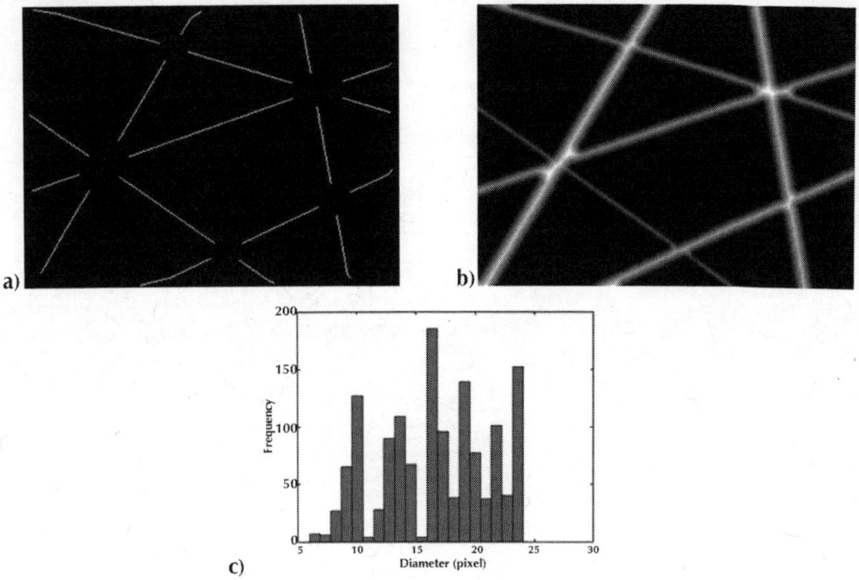

Figure 7.10. New distance transform method: (a) the skeleton of the simple simulated image shown in Fig. 7.6a after deleting the intersection points, (b) the distance map, (c) histogram of fiber diameter distribution.

The procedure for determining fiber diameter via this approach is summarized in Fig. 7.11. The method is efficient, reliable, accurate, and so fast and has the capability of being used as an online method for quality control.

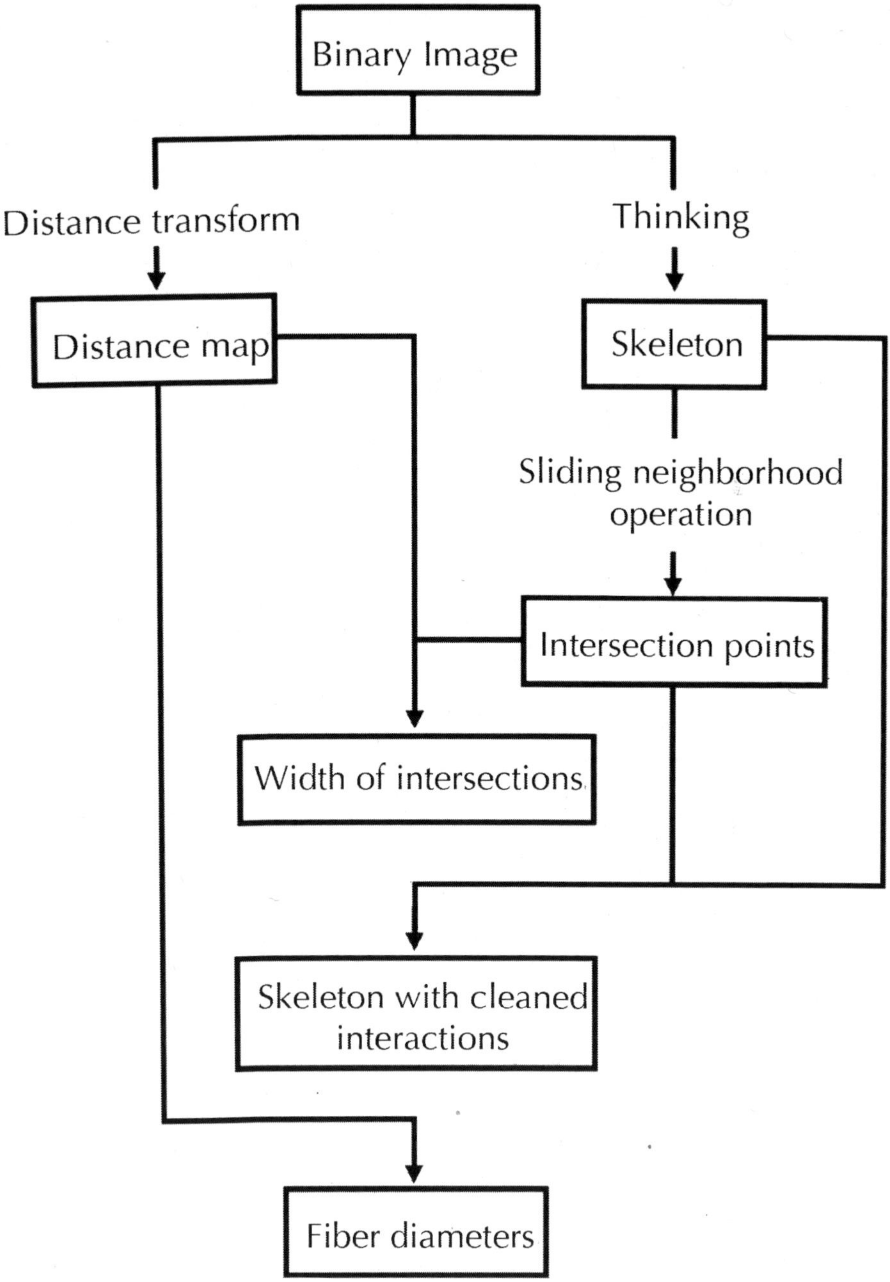

Figure 7.11. Flowchart of the new distance transform method.

Validation of Methods

A geometric model has been considered here to simulate electrospun fiberwebs. There are three widely used methods for generating random network of lines. These are called S-randomness, μ-randomness (suitable for generating a web of continuous filaments), and I-randomness (suitable for generating a web of staple fibers). Under this scheme, a line with a specified thickness is defined by the perpendicular distance d from a fixed reference point O located in the center of the image and the angular position of the perpendicular α. Distance d is limited to the diagonal of the image. Figure 7.12 demonstrates this procedure.

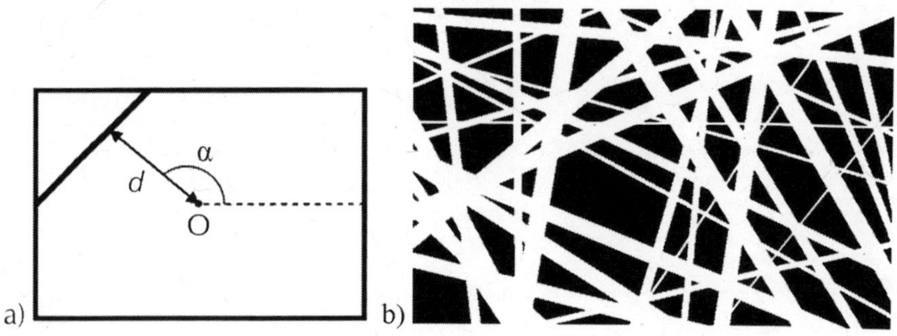

Figure 7.12. μ-randomness: (a) schematic view of the procedure, (b) a typical simulated image generated using this approach.

One of the most important features of simulation is that it allows several structural characteristics to be taken into consideration with the simulation parameters. These parameters are: web density (controlled as line density), angular density (sampled from a normal or random distribution), distance from the reference point (sampled from a random distribution), line thickness (sampled from a normal distribution), and image size.

Thresholding

Global thresholding, however, is very sensitive to any inhomogeneities in the gray-level distributions of object and background pixels. Figure 7.13a illustrates a typical micrograph obtained from electron microscopy. As it is shown in Fig. **7.**13b, global thresholding resulted in some broken fiber segments. In order to eliminate the effect of inhomogeneities, *local thresholding* scheme could be used. In this approach, the image is divided into subimages where the inhomogeneities are negligible. Then, optimal thresholds are found for each subimage.

A common practice in this case is to use morphological *opening* to compensate for nonuniform background illumination. The morphological opening is a sequential application of an *erosion* operation followed by a *dilation* operation (i.e., opening = erosion + dilation) using the same *structuring element*. Dilation is an operation that grows or thickens objects in a binary image by adding pixels to the boundaries of objects. Erosion shrinks or thins objects in a binary image by removing pixels on object boundaries. The specific manner and extent of the thickening or thinning is controlled by the size and shape of the structuring element which is a matrix consisting of 0s and 1s having any arbitrary shape and size. Opening the image produces an estimate of the background provided large enough structuring element is used, so that it does not fit entirely within the objects (Fig. 7.13c). Subtracting the opened image from the original image, which is called *top-hat* transformation, results in an image with a reasonably even background (Fig. 7.13d). Now that the background is homogeneous and the edges of the objects are clearer, a global thresholding could be applied to provide the binary image. It could be shown that this process is equivalent to segment the image with locally varying thresholds.

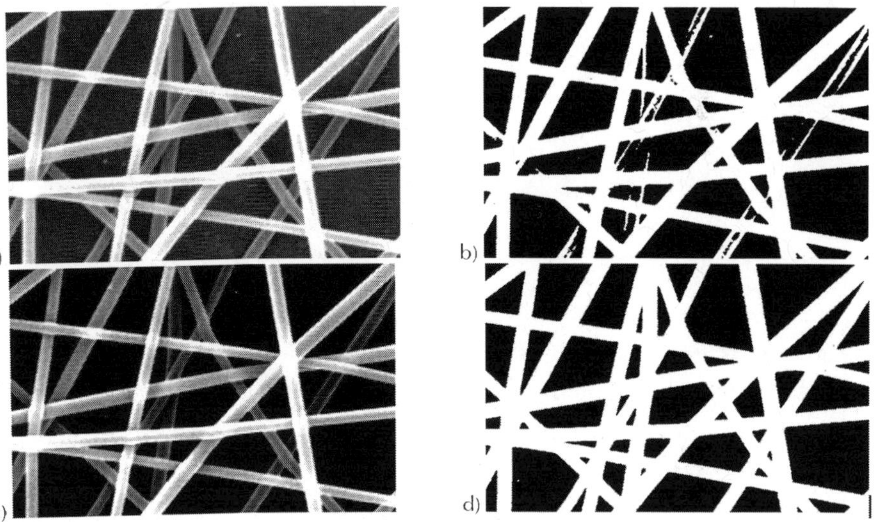

Figure 7.13. Thresholding: (a) a typical micrograph of electrospun web, (b) global thresholding, (c) top-hat transformation, (d) local thresholding.

KEYWORDS

- **Dilation**
- **Electrospun nanofiber webs**
- **Euclidean metric**
- **Fiber diameter**
- **Global thresholding**
- **Medial axis transformation**
- **Pruning procedure**

Chapter 8

Update on Control of Electrospun Web Pores Structure

INTRODUCTION

The electrospinning process uses high voltage to create an electric field between a droplet of polymer solution at the tip of a needle and a collector plate. When the electrostatic force overcomes the surface tension of the drop, a charged, continuous jet of polymer solution is ejected. As the solution moves away from the needle and toward the collector, the solvent evaporates and jet rapidly thins and dries. On the surface of the collector, a nonwoven web of randomly oriented solid nanofibers is deposited. Figure 8.1 illustrates the electrospinning setup.

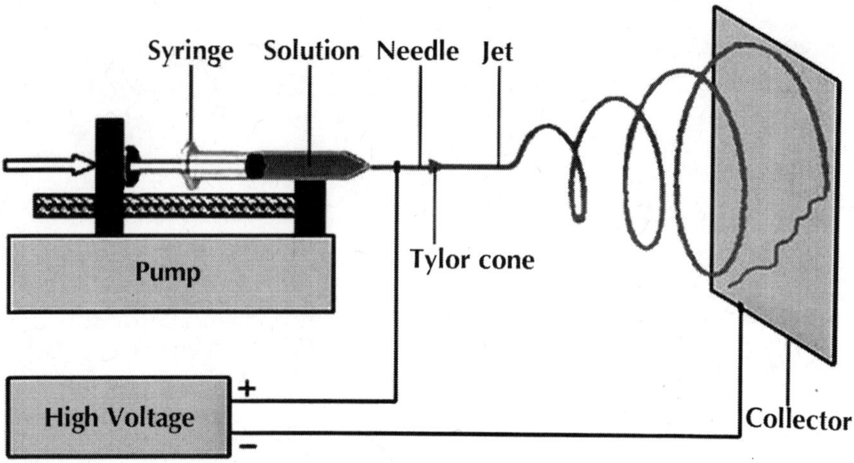

Figure 8.1. Electrospinning setup.

Material properties such as melting temperature and glass transition temperature as well as structural characteristics of nanofiber webs such as fiber diameter distribution, pore size distribution and fiber orientation distribution determine the physical and mechanical properties of the webs. The surface of electrospun fibers is important when considering end-use applications. For example, the ability to introduce porous surface features of a known size is required if nanoparticles need to be deposited on the surface of the fiber, if drug molecules are to be incorporated for controlled release, as tissue scaffolding materials and for acting as a cradle for enzymes. Besides, filtration performance of nanofibers is strongly related to their pore structure parameters, i.e., percent open area (POA) and pore-opening size distribution (PSD). Hence, the control of the pore of electrospun webs is of prime importance for the nanofibers that are being produced for these purposes. There is no literature available about the pore size and its distribution of electrospun fibers and in this chapter, the pore size and its distribution was measured using an image analysis technique.

METHODOLOGY

The porosity, ε_V, is defined as the percentage of the volume of the voids, V_v, to the total volume (voids plus constituent material), V_t, and is given by

$$\varepsilon_V = \frac{V_v}{V_t} \times 100 \tag{1}$$

Similarly, the POA, ε_A, that is defined as the percentage of the open area, A_o, to the total area A_t, is given by

$$\varepsilon_A = \frac{A_o}{A_t} \times 100 \tag{2}$$

Usually porosity is determined for materials with a three-dimensional structure, e.g., relatively thick nonwoven fabrics. Nevertheless, for two-dimensional textiles such as woven fabrics and relatively thin nonwovens it is often assumed that porosity and POA are equal.

The size of an individual opening can be defined as the surface area of the opening, although it is mostly indicated with a diameter called Equivalent Opening Size (EOS). EOS is not a single value, for each opening may differ. The common used term in this case is the diameter, O_i, corresponding with the equivalent circular area, A_i, of the opening.

$$O_i = (4A_i/\pi)^{1/2} \tag{3}$$

This diameter is greater than the side dimension of a square opening. A spherical particle with that diameter will never pass the opening (Fig. 8.2a) and may therefore not be considered as an equivalent dimension or equivalent diameter. This will only be possible if the diameter corresponds with the side of the square area (Fig. 8.2b). However, not all openings are squares, yet the equivalent square area of openings is used to determine their equivalent dimension because this simplified assumption results in one single opening size from the open area. It is the diameter of a spherical particle that can pass the equivalent square opening, hence the equivalent opening or pore size, O_i, results from

$$O_i = (A_i)^{1/2} \tag{4}$$

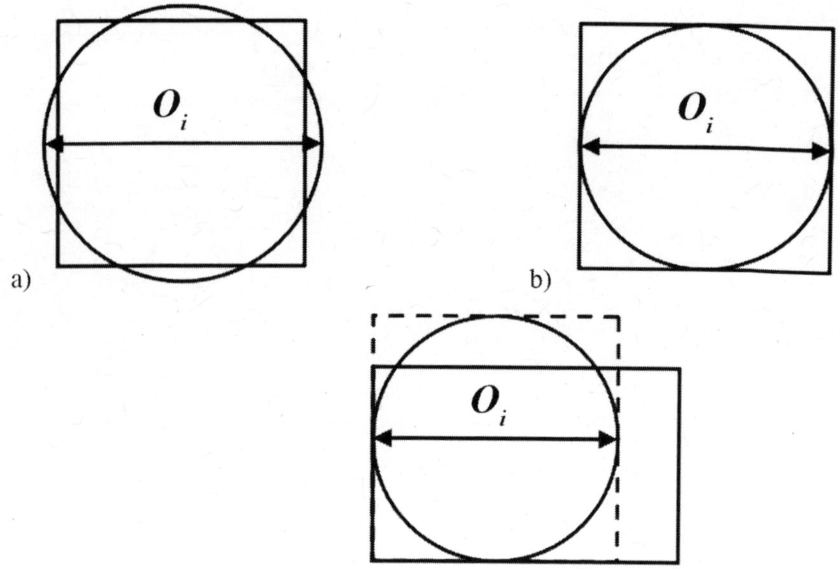

Figure 8.2. Equivalent opening size, Oi, based on (a) equivalent area, (b) equivalent size.

From the EOSs, PSD and an equivalent diameter for which a certain percentage of the opening have a smaller diameter (O_x, pore opening size that x percent of pores are smaller than that size) may be measured.

The PSD curves can be used to determine the uniformity coefficient, C_u, of the investigated materials. The uniformity coefficient is a measure for the uniformity of the openings and is given by

$$C_u = O_{60} / O_{10} \tag{5}$$

The ratio equals 1 for uniform openings and increases with decreasing uniformity of the openings .

Pore characteristic is one of the main tools for evaluating the performance of any nonwoven fabric and for electrospun webs as well. Understanding the link between processing parameters and pore structure parameters will allow for better control over the properties of electrospun fibers. Therefore there is a need for the design of nanofibers to meet specific application needs. Various techniques may be used to evaluate pore characteristics of porous materials including sieving techniques (dry, wet and hydrodynamic sieving), mercury porosimetry and flow porosimetry (bubble point method). As one goes about selecting a suitable technique for characterization, the associated virtues and pitfalls of each technique should be examined. The most attractive option is a single technique which is non-destructive, yet capable of providing a comprehensive set of data .

Mercury Porosimetry

Mercury porosimetry is a well known method which is often used to study porous materials. This technique is based on the fact that mercury as a non-wetting liquid does not intrude into pore spaces except under applying sufficient pressure. Therefore, a relationship can be found between the size of pores and the pressure applied.

In this method, a porous material is completely surrounded by mercury and pressure is applied to force the mercury into pores. As mercury pressure increases the large pores are filled with mercury first. Pore sizes are calculated as the mercury pressure increases. At higher pressures, mercury intrudes into the fine pores and when the pressure reaches a maximum, total open pore volume and porosity are calculated.

The mercury porosimetry thus gives a PSD based on total pore volume and gives no information regarding the number of pores of a porous material. Pore sizes ranging from 0.0018 to 400 μm can be studied using mercury porosimetry. Pore sizes smaller than 0.0018 μm are not intruded with mercury and this is a source of error for porosity and PSD calculations. Furthermore, mercury porosimetry does not account for closed pores as mercury does not intrude into them. Due to applying high pressures, sample collapse and compression is possible, hence it is not suitable for fragile compressible materials such as nanofiber sheets. Other concerns would include the fact that it is assumed that the pores are cylindrical, which is not the case in reality. After the mercury intrusion test, sample decontamination at specialized facilities is required as the highly toxic mercury is trapped within the pores. Therefore this dangerous and destructive test can only be performed in well-equipped labs .

Image Analysis

Because of its convenience to detect individual pores in a nonwoven image, it seemed to be advantageous to use image analysis techniques for pore measurement. Image analysis was used to measure pore characteristics of woven and nonwoven geotextiles.

Therefore, there is a need for developing an algorithm suitable for measuring the pore structure parameters in electrospun webs. In response to this need, a new image analysis based method has been developed which is presented in the following.

In this method, a binary image of the web is used as an input. First of all, voids connected to the image border are identified and cleared using morphological reconstruction where mask image is the input image and marker image is zero everywhere except along the border. Total area which is the number of pixels in the image is measured. Then the pores are labeled and each considered as an object. Here the number of pores may be obtained. In the next step, the number of pixels of each object as the area of that object is measured. Having the area of pores, the porosity and EOS regarding to each pore may be calculated. The data in pixels may then be converted to *nm*. Finally PSD curve is plotted and O_{50}, O_{95} and C_u are determined.

Real Webs: In order to measure pore characteristics of electrospun nanofibers using image analysis, images of the webs are required. These images called micrographs usually are obtained by Scanning Electron Microscope (SEM), Transmission Electron

Microscope (TEM) or Atomic Force Microscope (AFM). The images must be of high-quality and taken under appropriate magnifications.

The image analysis method for measuring pore characteristics requires the initial segmentation of the micrographs in order to produce binary images. This is a critical step because the segmentation affects the results dramatically. The typical way of producing a binary image from a grayscale image is by global thresholding where a single constant threshold is applied to segment the image. All pixels up to and equal to the threshold belong to object and the remaining belong to the background. One simple way to choose the threshold is picking different thresholds until one is found that produces a good result as judged by the observer. Global thresholding is very sensitive to any inhomogeneities in the gray-level distributions of object and background pixels. In order to eliminate the effect of inhomogeneities, local thresholding scheme could be used. In this approach, the image is divided into subimages where the inhomogeneities are negligible. Then optimal thresholds are found for each subimage. A common practice in this case, which is used in this contribution, is to preprocess the image to compensate for the illumination problems and then apply a global thresholding to the preprocessed image. It can be shown that this process is equivalent to segment the image with locally varying thresholds. In order to automatically select the appropriate thresholds, a new method is employed. This method chooses the threshold to minimize intraclass variance of the black and white pixels. As it is shown in Fig. 8.3, global thresholding resulted in some broken fiber segments. This problem was solved using local thresholding.

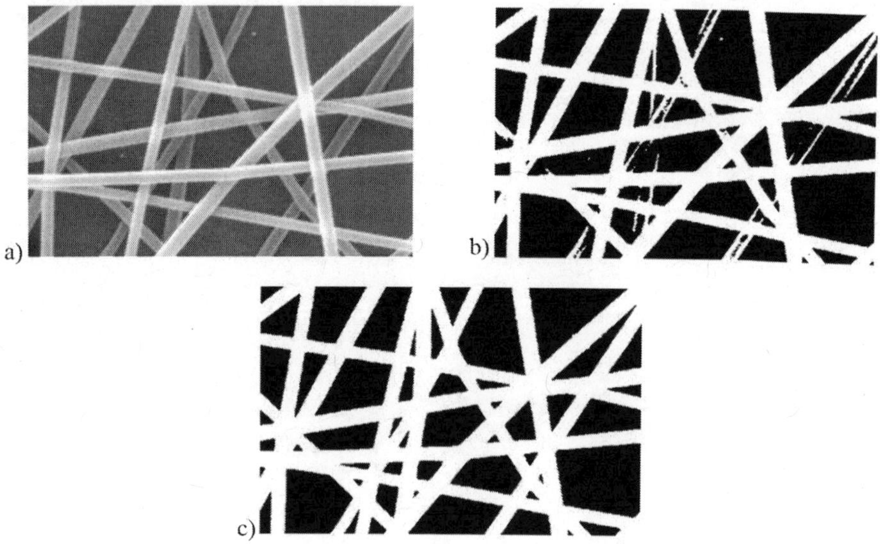

Figure 8.3. (a) A real web, (b) Global thresholding, (c) Local thresholding.

Note that, since the process is extremely sensitive to noise contained in the image, preceding segmentation, a procedure to clean the noise and enhance the contrast of the image is necessary.

Simulated Webs: It is known that the pore characteristics of nonwoven webs are influenced by web properties and so are those of electrospun webs. There are no reliable models available for predicting these characteristics as a function of web properties. In order to explore the effects of some parameters on pore characteristics of electrospun nanofibers, simulated webs are generated. These webs are images simulated by straight lines. There are three widely used methods for generating random network of lines. These are called S-randomness, μ-randomness (suitable for generating a web of continuous filaments) and I-randomness (suitable for generating a web of staple fibers). In this chapter, μ-randomness procedure for generating simulated images was used. Under this scheme, a line with a specified thickness is defined by the perpendicular distance d from a fixed reference point O located in the center of the image and the angular position of the perpendicular α. Distance d is limited to the diagonal of the image. Fig. 8.4 demonstrates this procedure.

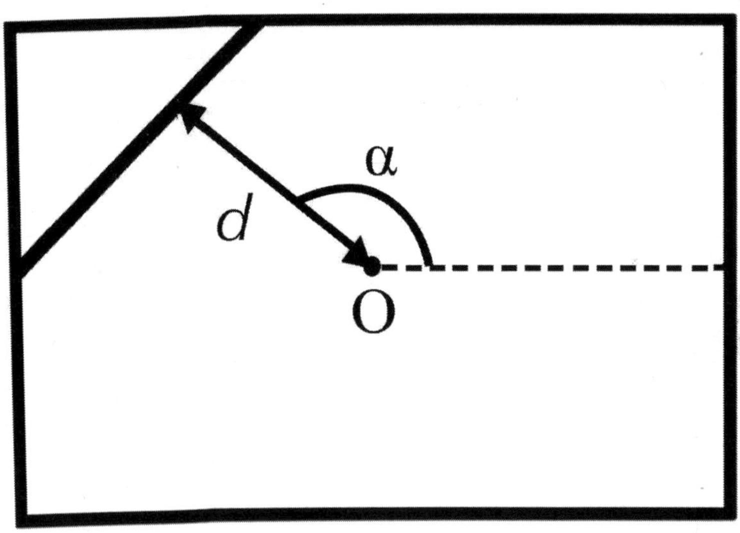

Figure 8.4. Procedure for μ-randomness.

One of the most important features of simulation is that it allows several structural characteristics to be taken into consideration with the simulation parameters. These parameters are: web density (controlled as line density), angular density (sampled from a normal or random distribution), distance from the reference point (sampled from a random distribution), line thickness (sampled from a normal distribution) and image size.

EXPERIMENT

Nanofiber webs were obtained from electrospinning of PVA with average molecular weight of 72,000 g/mol (MERCK) at different processing parameters for attaining different pore characteristics. Table 8.1 summarizes the electrospinning parameters used for preparing the webs. The micrographs of the webs were obtained using Philips (XL-30) environmental SEM under magnification of 10,000X after being gold coated. Figure 8.5 shows the micrographs of electrospun webs.

Table 8.1. Electrospinning parameters used for preparing nanofiber webs.

No.	Concentration (%)	Spinning Distance (Cm)	Voltage (KV)	Flow Rate (ml/h)
1	8	15	20	0.4
2	12	20	15	0.2
3	8	15	20	0.2
4	8	10	15	0.3
5	10	10	15	0.2

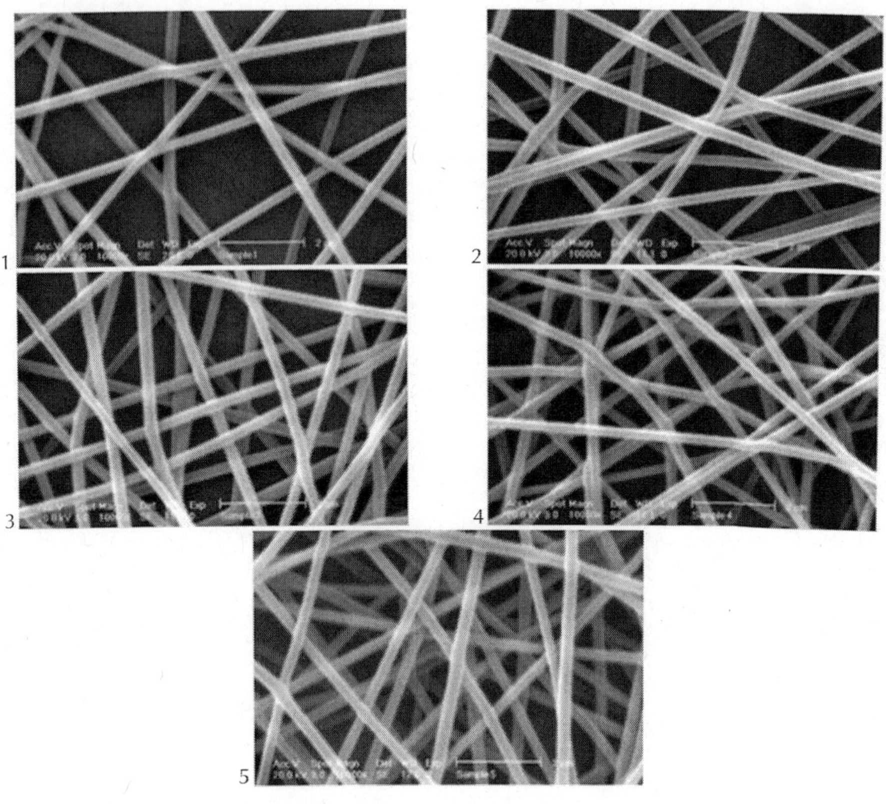

Figure 8.5. Micrographs of the electrospun webs.

RESULTS AND DISCUSSION

Due to previously mentioned reasons, sieving methods and mercury porosimetry are not applicable for measuring pore structure parameters in nano-scale. The only method which seems to be practical is flow porosimetry. However, since in this contribution, the nanofibers were made of PVA, finding an appropriate liquid for the test to be performed is almost impossible because of solubility of PVA in both organic and inorganic liquids.

As an alternative, image analysis was employed to measure pore structure parameters in electrospun nanofiber webs. PSD curves of the webs, determined using the image analysis method, are shown in Fig. 8.6. Pore characteristics of the webs (O_{50}, O_{95}, C_u, number of pores, porosity) measured by this method are presented in Table 8.2. It is seen that decreasing the porosity, O_{50} and O_{95} decrease. C_u also decreases with respect to porosity, that is to say increasing the uniformity of the pores. Number of pores has an increasing trend with decreasing the porosity.

Table 8.2. Pore characteristics of electrospun webs.

No.	O_{50}		O_{95}		C_u	Pore No.	Porosity
	pixel	Nm	pixel	nm			
1	39.28	513.9	94.56	1237.1	8.43	31	48.64
2	27.87	364.7	87.66	1146.8	5.92	38	34.57
3	26.94	352.5	64.01	837.4	3.73	64	26.71
4	22.09	289.0	60.75	794.8	3.68	73	24.45
5	19.26	252.0	44.03	576.1	2.73	69	15.74

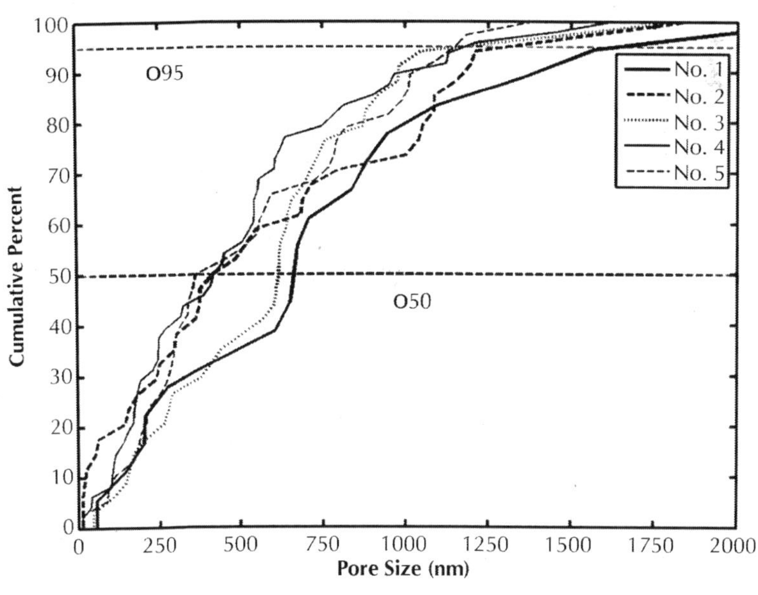

Figure 8.6. PSD curves of electrospun webs.

The image analysis method presents valuable and comprehensive information regarding to pore structure parameters in nanofiber webs. This information may be exploited in preparing the webs with needed pore characteristics to use in filtration, biomedical applications, nanoparticle deposition and other purposes. The advantages of the method are listed below:

1. The method is capable of measuring pore structure parameters in any nanofiber webs with any pore features and it is applicable even when other methods may not be employed.
2. It is so fast. It takes less than a second for an image to be analyzed (using GHz processor).
3. The method is direct and so simple. Pore characteristics are measured from the area of the pores which is defined as the number of pixels of the pores.
4. There is no systematic error in measurement (such as assuming pores to be cylindrical in mercury and flow porosimetry and the errors associated with the sieving methods which were mentioned). Once the segmentation is successful, the pore sizes will be measured accurately. The quality of images affects the segmentation procedure. High-quality images reduce the possibility of poor segmentation and enhance the accuracy of the results.
5. It gives a complete PSD curve.
6. There is no cost involved in the method and minimal technical equipments are needed (SEM for obtaining the micrographs of the samples and a computer for analysis).
7. It has the capability of being used as an on-line quality control technique for large scale production.
8. The results obtained by image analysis are reproducible.
9. It is not a destructive method. A very small amount of sample is required for measurement.

In an attempt to establish the effects of some structural properties on pore characteristics of electrospun nanofibers, two sets of simulated images with varying properties were generated. The simulated images reveal the degree to which fiber diameter and density affect the pore structure parameters. The first set contained images with the same density varying in fiber diameter and images with the same fiber diameter varying in density. Each image had a constant diameter. The second set contained images with the same density and mean fiber diameter while the standard deviation of fiber diameter varied. The details are given in Tables 8.3 and 8.4. Typical images are shown in Figs. 8.7 and 8.8.

Table 8.3. Structural characteristics of first set images.

No.	Angular Range	Line Density	Line Thickness
1	0–360	20	5
2	0–360	30	5

Table 8.3. *(Continued)*

No.	Angular Range	Line Density	Line Thickness
3	0–360	40	5
4	0–360	20	10
5	0–360	30	10
6	0–360	40	10
7	0–360	20	20
8	0–360	30	20
9	0–360	40	20

Table 8.4. Structural characteristics of second set images.

No.	Angular Range	Line Density	Line Thickness	
			Mean	Std
1	0–360	30	15	0
2	0–360	30	15	4
3	0–360	30	15	8
4	0–360	30	15	10

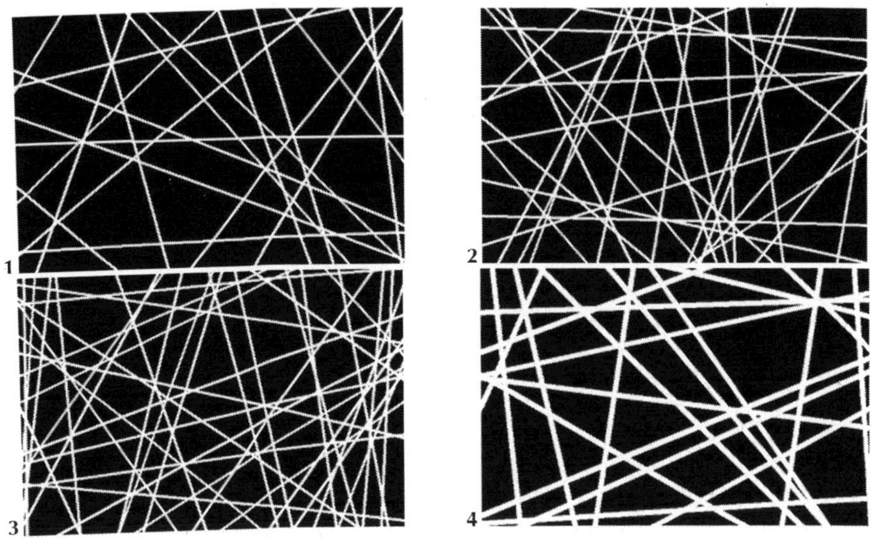

Figure 8.7. *(Continued)*

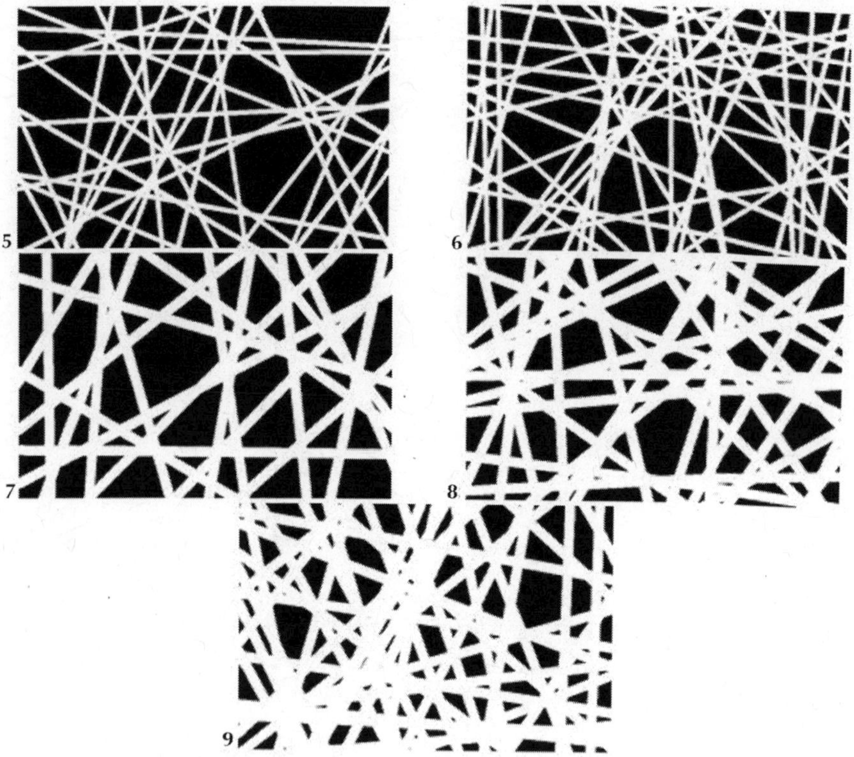

Figure 8.7. Simulated images of the first set.

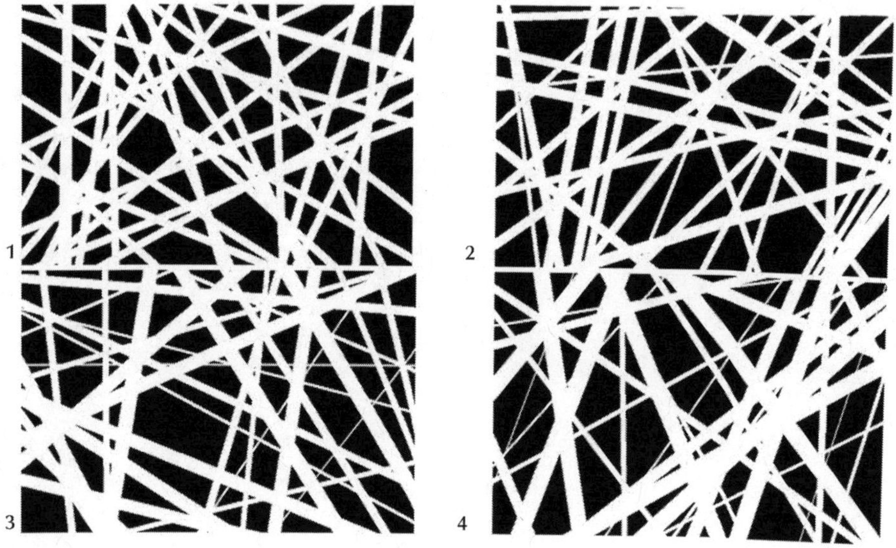

Figure 8.8. Simulated images of the second set.

Pore structure parameters of the simulated webs were measured using image analysis method. Table 8.5 summarizes the pore characteristics of the simulated images in the first set. For the webs with the same density, increasing fiber diameter resulted in a decrease in O_{95}, number of pores and porosity. Assuming the web density to be constant, increasing fiber diameter, the ratio of area of fibers to total area (that's to say, the proportion of white pixels to total pixels in the image) increases, reducing the porosity. It could be imagined that as the fibers get thicker, small pores are covered with the fibers, lowering the number of pores. An increase of fiber diameter at a given web density, results in smaller pores; hence O_{95} decreases. No particular trends were observed for O_{50} and C_u. In the case of O_{50}, it is because the effect of fiber diameter is more significant on larger pores while O_{50} is related to mostly small pores and there seems to be other parameters such as the arrangement of the fibers which influence O_{50} more significantly rather than fiber diameter. Since in Eq. (5), O_{10} is in the denominator of the fraction, C_u is very sensitive to variation of O_{10}. This is while O_{10} tends to vary much and almost regardless of fiber diameter (due to aforementioned reason since it is related to very small pores). Hence, other factors e.g., the way fibers arrange are more dominant and C_u varies regardless of fiber diameter.

Table 8.5. Pore characteristics of the first set of simulated images.

No.	O_{50}	O_{95}	C_u	Pore No.	Porosity
1	27.18	100.13	38.38	84	79.91
2	15.52	67.31	22.20	182	71.78
3	13.78	52.32	18.71	308	69.89
4	36.65	94.31	43.71	67	66.10
5	17.89	61.64	22.67	144	53.67
6	12.41	51.60	16.70	245	47.87
7	24.49	86.90	33.11	58	41.05
8	16.31	56.07	21.66	108	32.53
9	13.11	45.38	17.75	126	22.01

Figures 8.9 and 8.10 show the PSD curves of the simulated images in the first set. As the web density increases, the effects of fiber diameter are less pronounced since the PSD curves of the webs become closer to each other.

For the webs with the same fiber diameter, increasing the density resulted in a decrease in O_{50}, O_{95}, C_u and porosity whereas number of pores increased with the density. For the same fiber diameter, total number of fibers and indeed total number of crossovers increases as web density raises; suggesting more number of pores. It is quite trivial that at a given fiber diameter, the ratio of area of fibers to total area increases as the webs get denser; thus lowering the porosity. Increasing the web density leads to more number of crossovers. Therefore large pores are split into several smaller pores. As a result, O_{50} and O_{95} decrease. Furthermore, this fracture of the pores results in a less variation of the pore size. Hence, uniformity increases; that's to say, C_u decreases.

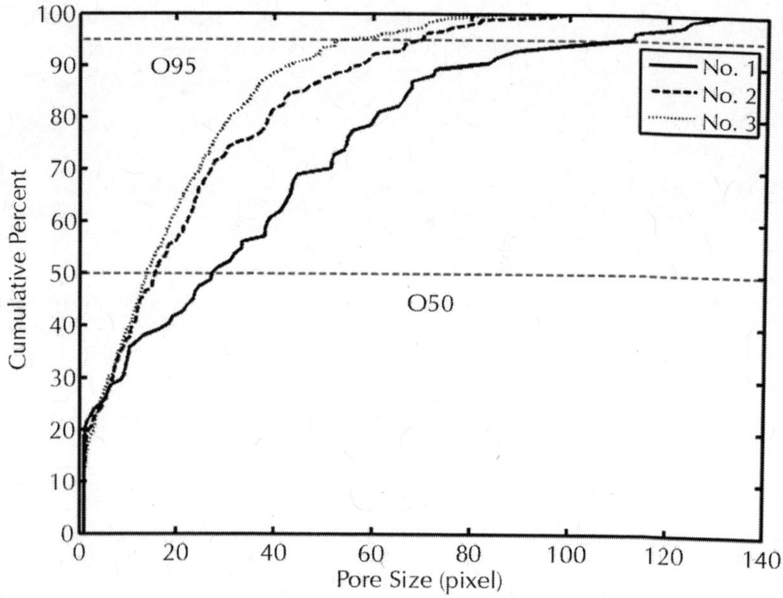

a)

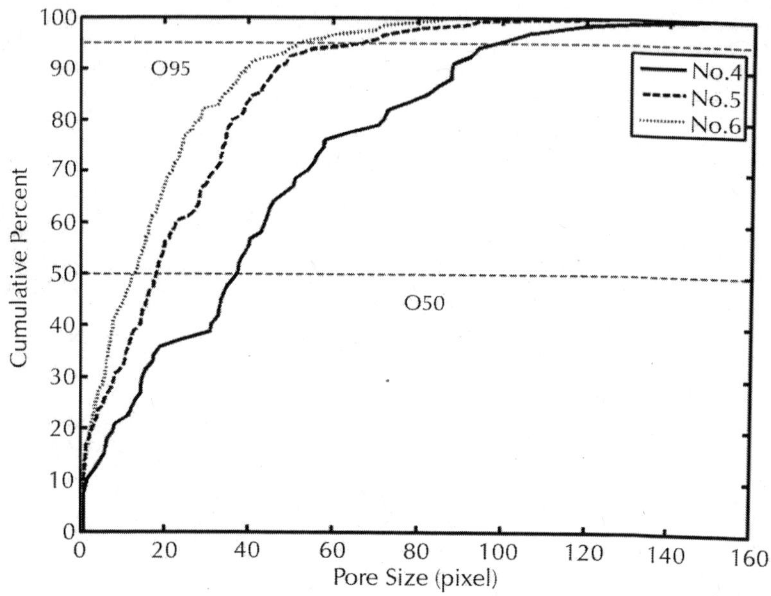

b)

Figure 8.9. *(Continued)*

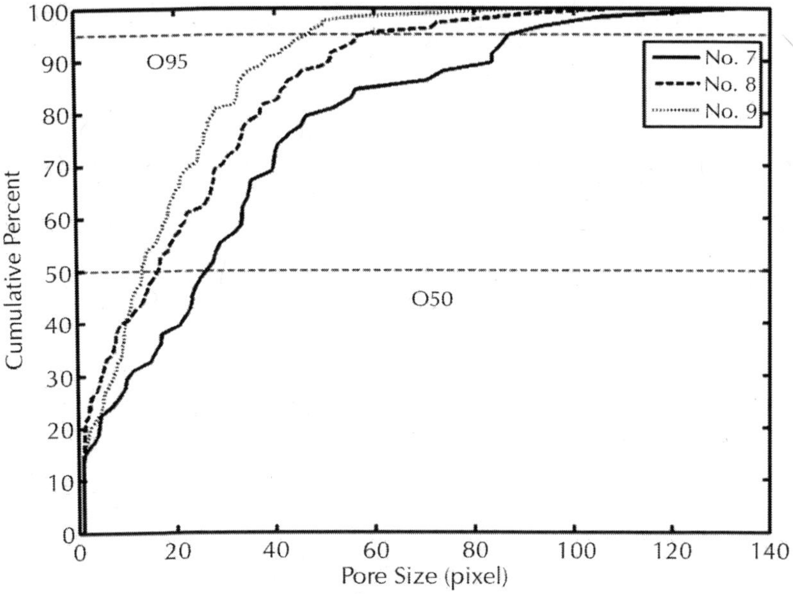

c)

Figure 8.9. PSD curves of the first set of simulated images; effect of density, images with the diameter of (a) 5, (b) 10, (c) 20 pixels.

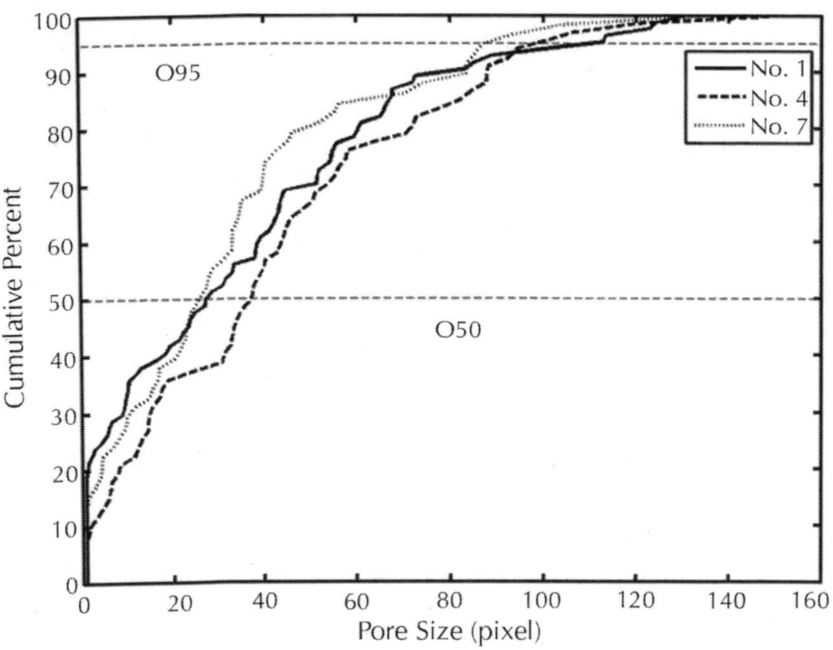

Figure 8.10. *(Continued)*

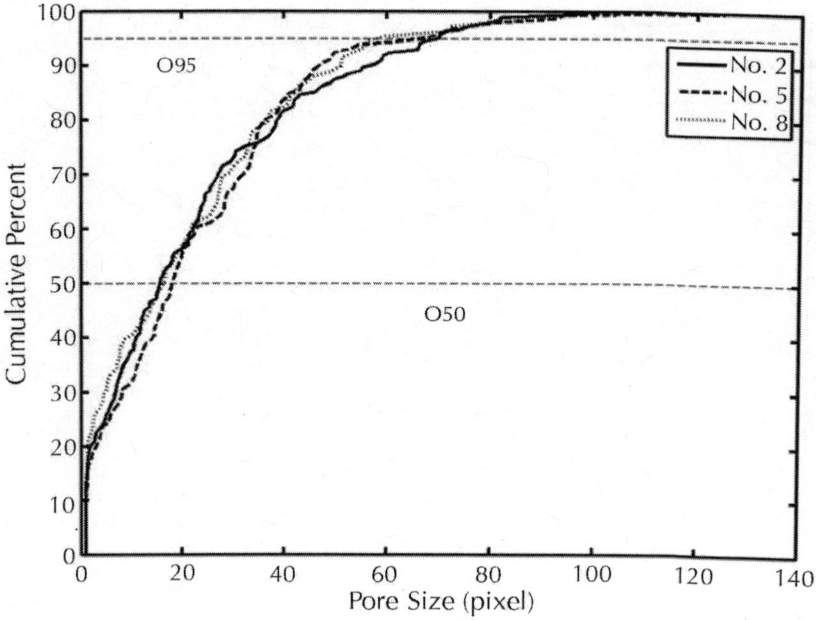

b)

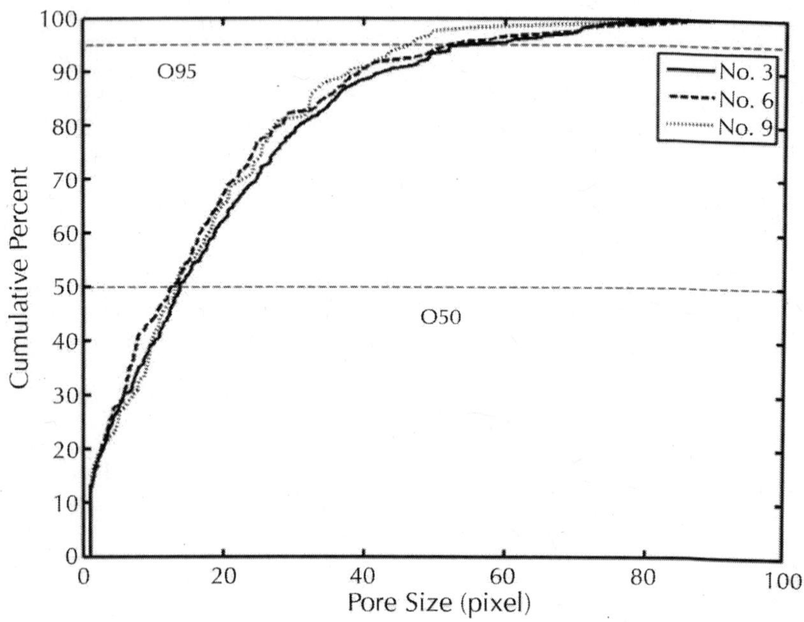

c)

Figure 8.10. PSD curves of the first set of simulated images; effect of fiber diameter, images with the density of (a) 20, (b) 30, (c) 40 lines.

Table 8.6 summarizes the pore characteristics of the simulated images in the second set. No significant effects for variation of fiber diameter on pore characteristics were observed. Suggesting that average fiber diameter is determining factor not variation of diameter. Figure 8.11 shows the PSD curves of the simulated images in the second set. Holding web density and average fiber diameter constant, the ratio of area of fibers to total area remains the same or fluctuates mostly due to arrangement of the fibers and regardless of variation of fiber diameter. As a result, porosity is not related to variation of fiber diameter. No trends in O_{50} and O_{95} with respect to variation of fiber diameter were observed. This could be attributed to different pore sizes regarding to how thin and thick fibers arrange. The changes in number of pores seem to be independent of variation of fiber diameter as well. It could also be attributed to the arrangement of the fibers.

Table 8.6. Pore characteristics of the second set of simulated images.

No.	O_{50}	O_{95}	C_u	Pore No.	Porosity
1	14.18	53.56	18.79	133	35.73
2	13.38	61.66	20.15	136	41.89
3	18.14	59.35	22.07	121	41.03
4	15.59	62.71	20.20	112	37.77

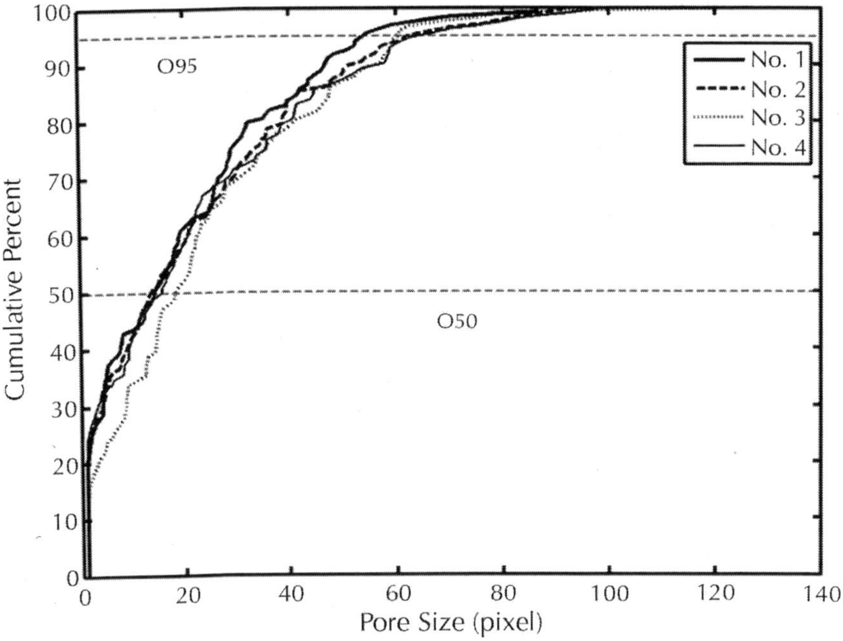

Figure 8.11. PSD curves of the second set of simulated images, the effect of fiber diameter variation.

KEYWORDS

- **Equivalent opening size**
- **Global thresholding**
- **Mercury porosimetry**
- **Pore-opening size distribution (PSD) curves**

Chapter 9

Update on Control of some of the Governing Parameters in Electrospinning

INTRODUCTION

In this chapter, response surface methodology (RSM) was employed to investigate the simultaneous effects of four of the most important parameters, in electrospinning of polyvinyl alcohol (PVA) nanofibers.

Figure 9.1 shows a schematic illustration of electrospinning setup.

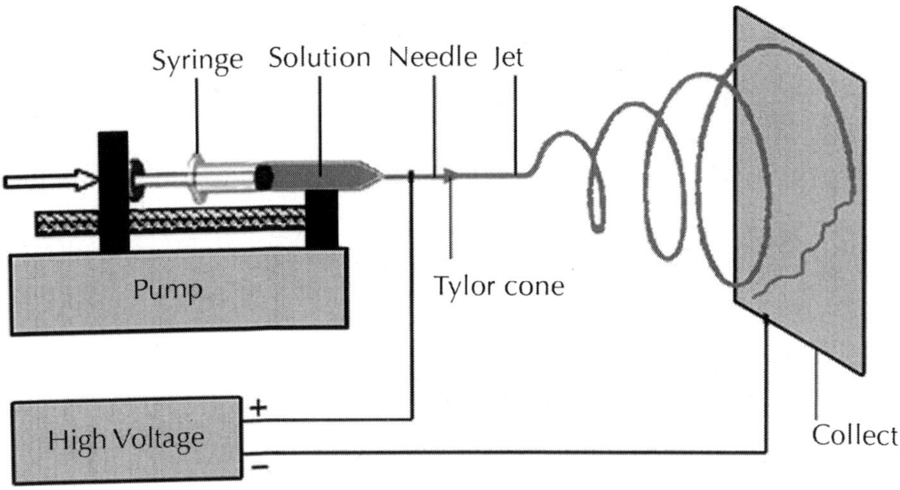

Figure 9.1. A typical image of electrospinning process.

According to various outstanding properties such as very small fiber diameters, large surface area per mass ratio, high porosity along with small pore sizes, flexibility, and superior mechanical properties, electrospun nanofiber mats have found numerous applications in diverse areas. For example, in biomedical field nanofibers plays a substantional role in tissue engineering, drug delivery, and wound dressing. Moreover, the use of nanofibers in protective clothing, filtration technology, and reinforcement of composite materials is extremely significant for developing of specific products by manipulation of materials in nanoscales. In the mean time, those applications related to micro-electronics like battery, supercapacitors, transistors, sensors, and display devices.

The physical characteristics of electrospun nanofibers such as fiber diameter depend on various parameters which are mainly divided into three categories: solution properties (solution viscosity, solution concentration, polymer molecular weight, and surface tension), processing conditions (applied voltage, volume flow rate, spinning distance, and needle diameter), and ambient conditions (temperature, humidity, and atmosphere pressure) . Numerous applications require nanofibers with desired properties suggesting the importance of the process control.

Affecting characteristics of the final product such as physical, mechanical, and electrical properties, fiber diameter is one of the most important structural features in electrospun nanofiber mats. It should be noted that filters composed of fibers with smaller diameters have higher filtration efficiencies. It is also reported that sensitivity of sensors increase with decreasing the mean fiber diameter due to the higher surface area. It was also shown that in polymer batteries consisting of electrospun polyvinylidene fluoride (PVdF) fibrous electrolyte, lower mean fiber diameter results in a higher electrolyte uptake thereby increasing ionic conductivity of the mat. Furthermore, fiber diameters of electrospun polyethylene oxide terephthalate/polybutylene terephthalate (PEOT/PBT) scaffolds influencing on cell seeding, attachment, and proliferation. They also studied the release of dye incorporated in electrospun scaffolds and observed that with increasing fiber diameter, the cumulative release of the dye (methylene blue) decreased. Carbonization and activation conditions as well as the structure and properties of the ultimate carbon fibers are also affected by the diameters of the precursor polyacrylonitril (PAN) nanofibers. Consequently, precise control of the electrospun fiber diameter is very crucial.

A few techniques such as orthogonal experimental design and using power law relationships have been reported in the literature for quantitative study of electrospun nanofiber. However, researchers mostly paid attention to RSM technique due to its simplicity and its ability to take into account the interactions between the parameters. Researchers employed RSM to model mean fiber diameter of electrospun regenerated Bombyx mori silk with electric field and concentration at two spinning distances. They applied a full factorial experimental design at three levels of each parameter leading to nine treatments of factors and used a quadratic polynomial to establish a relationship between mean fiber diameter and the variables. Increasing the concentration at constant electric field resulted in an increase in mean fiber diameter. Different impacts for the electric field were observed depending on solution concentration. Since the effects of solution concentration and electric field strength on mean fiber diameter changed at different spinning distances, they suggested that some interactions and coupling effects are present between the parameters.

Researchers also exploited the RSM for quantitative study of PAN and poly D, L-lactide (PDLA) respectively. The only difference observed in the procedure was the use of four levels of concentration in the former case. They included the standard deviation of fiber diameter in their investigations by which they were able to provide additional information regarding the morphology of electrospun nanofibers and its variations at different conditions. In the most recent investigation in this field, Yördem, researchers utilized RSM to correlate mean and coefficient of variation (CV) of electrospun PAN nanofibers to solution concentration and applied voltage at three different spinning distances. They employed a face-centered central composite design

(FCCD) along with a full factorial design at two levels resulting in 13 treatments at each spinning distance. A cubic polynomial was then used to fit the data in each case. As previous studies, fiber diameter was very sensitive to changes in solution concentration. Voltage effect was more significant at higher concentrations demonstrating the interaction between parameters.

According to studies, there are some interactions between electrospinning parameters. In this chapter for the first time, the simultaneous effects of four electrospinning parameters (solution concentration, spinning distance, applied voltage, and volume flow rate) on mean and standard deviation of electrospun PVA fiber diameter were systematically investigated.

EXPERIMENT

Solution Preparation and Electrospinning

PVA with molecular weight of 72,000 *g/mol* and degree of hydrolysis of >98% was obtained from Merck and used as received. Distilled water as solvent was added to a predetermined amount of PVA powder to obtain 20 *ml* of solution with desired concentration. The solution was prepared at 80°C and gently stirred for 30 min to expedite the dissolution. After the PVA had completely dissolved, the solution was transferred to a 5 *ml* syringe and became ready for spinning of nanofibers. The experiments were carried out on a horizontal electrospinning setup shown schematically in Figure 9.1. The syringe containing PVA solution was placed on a syringe pump (New Era NE-100) used to dispense the solution at a controlled rate. A high voltage DC power supply (Gamma High Voltage ES-30) was used to generate the electric field needed for electrospinning. The positive electrode of the high voltage supply was attached to the syringe needle via an alligator clip and the grounding electrode was connected to a flat collector wrapped with aluminum foil where electrospun nanofibers were accumulated to form a nonwoven mat. The electrospinning was carried out at room temperature. Subsequently, the aluminum foil was removed from the collector. A small piece of mat was placed on the sample holder and gold sputter-coated (Bal-Tec). Thereafter, the micrograph of electrospun PVA fibers was obtained using scanning electron microscope (SEM, Phillips XL-30) under magnification of 10,000X. Quite recently, the authors established a couple of image analysis based techniques entitled as *direct tracking* and *new distance transform* for measuring electrospun nanofiber diameter. In this chapter, fiber diameter distribution for each specimen was determined from the SEM micrograph by new distance transform method due to its effectiveness.

Choice of Parameters and Range

Solution concentration (C), spinning distance (d), applied voltage (V), and volume flow rate (Q) were selected to be the most influential parameters. The next step is to choose the ranges over which these factors are varied. Process knowledge, which is a combination of practical experience and theoretical understanding, is required to fulfill this step. The aim is here to find an appropriate range for each parameter where dry, bead-free, stable, and continuous fibers without breaking up to droplets are obtained. This goal could be achieved by conducting a set of preliminary experiments while having the previous works in mind along with utilizing the reported relationships.

The relationship between intrinsic viscosity ($[\eta]$) and molecular weight (M) is given by the well-known Mark–Houwink–Sakurada equation as follows:

$$[\eta] = KM^{a} \tag{1}$$

where K and a are constants for a particular polymer-solvent pair at a given temperature. For the PVA with molecular weight in the range of 69,000 $g/mol < M < 690{,}000$ g/mol in water at room temperature, $K = 6.51$ and $a = 0.628$ were found by many researchers. Using these constants in the equation, the intrinsic viscosity for PVA in this chapter (molecular weight of 72,000 g/mol) was calculated to be $[\eta] = 0.73$.

Polymer chain entanglements in a solution can be expressed in terms of Berry number (B), which is a dimensionless parameter and defined as the product of intrinsic viscosity and polymer concentration ($B=[\eta]C$). For each molecular weight, there is a minimum concentration at which the polymer solution cannot be electrospun. It is also observed that $B>5$ is required to form stabilized fibrous structures in electrospinning of PVA. On the other hand, they reported the formation of flat fibers at $B>9$. Therefore, the appropriate range in this case could be found within $5<B<9$ domain which is equivalent to $6.8\%<C<12.3\%$ in terms of concentration of PVA. Furthermore, it is observed that beaded fibers were electrospun at low solution concentration. Hence, it was thought that the domain $8\% \leq C \leq 12\%$ would warrant the formation of stabilized bead-free fibers with circular cross-sections. This domain was later justified by performing some preliminary experiments.

As for determining the appropriate range of applied voltage, referring to previous works, it was observed that the changes of voltage lay between 5 kV and 25 kV depending on experimental conditions; voltages above 25 kV were rarely used. Afterwards, a series of experiments were carried out to obtain the desired voltage domain. At $V<10$ kV, the voltage was too low to spin fibers and $10\ kV \leq V < 15\ kV$ resulted in formation of fibers and droplets; in addition, electrospinning was impeded at high concentrations. In this regard, $15\ kV \leq V \leq 25\ kV$ was selected to be the desired domain for applied voltage.

The use of 5 – 20 cm for spinning distance was reported in the literature. Short distances are suitable for highly evaporative solvents whereas it results in wet coagulated fibers for nonvolatile solvents due to insufficient evaporation time. Since water was used as solvent for PVA in this chapter, short spinning distances were not expected to be favorable for dry fiber formation. Afterwards, this was proved by experimental observations and $10\ cm \leq d \leq 20\ cm$ was considered as the effective range for spinning distance.

Few researchers have addressed the effect of volume flow rate. Therefore in this case, the attention was focused on experimental observations. At $Q<0.2$ ml/h, in most cases especially at high polymer concentrations, the fiber formation was hindered due to insufficient supply of solution to the tip of the syringe needle. Whereas, excessive feed of solution at $Q>0.4$ ml/h incurred formation of droplets along with fibers. As a result, $0.2\ ml/h \leq Q \leq 0.4\ ml/h$ was chosen as the favorable range of flow rate in this chapter.

Experimental Design

Consider a process in which several factors affect a response of the system. In this case, a conventional strategy of experimentation, which is extensively used in practice, is the one-factor-at-a-time approach. The major disadvantage of this approach is its failure to consider any possible interaction between the factors, say the failure of one factor to produce the same effect on the response at different levels of another factor. For instance, suppose that two factors A and B affect a response. At one level of A, increasing B causes the response to increase, while at the other level of A, the effect of B totally reverses and the response decreases with increasing B. As interactions exist between electrospinning parameters, this approach may not be an appropriate choice for the case which is concerned in the present contribution. The correct strategy to deal with several factors is to use a full factorial design. In this method, factors are all varied together; therefore all possible combinations of the levels of the factors are investigated. This approach is very efficient, makes the most use of the experimental data and takes into account the interactions between factors.

It is trivial that in order to draw a line at least two points and for a quadratic curve at least three points are required. Hence, three levels were selected for each parameter so that it would be possible to use quadratic models. These levels were chosen equally spaced. A full factorial experimental design with four factors (solution concentration, spinning distance, applied voltage, and flow rate) each at three levels (3^4 design) were employed resulting in 81 treatment combinations. This design is shown in Fig. 9.2.

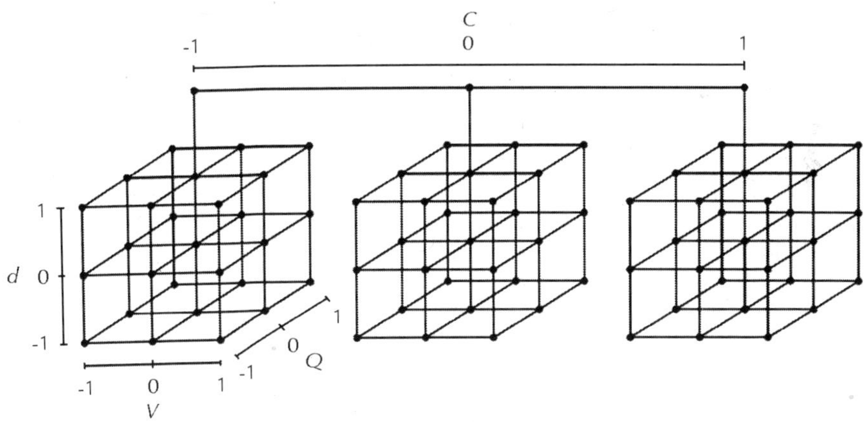

Figure 9.2. 3 Full factorial experimental design used in this contribution.

−1, 0, and 1 are coded variables corresponding to low, intermediate and high levels of each factor respectively. The coded variables (x_j) were calculated using Eq. (2) from natural variables (ξ_i). The indices 1 to 4 represent solution concentration, spinning distance, applied voltage, and flow rate respectively. In addition to experimental data, 15 treatments inside the design space were selected as test data and used for evaluation of the models.

$$x_j = \frac{\xi_j - [\xi_{hj} + \xi_{lj}]/2}{[\xi_{hj} - \xi_{lj}]/2} \tag{2}$$

Response Surface Methodology

RSM is a combination of mathematical and statistical techniques useful for empirical modeling and analysis of such systems. The application of RSM is in situations where several input variables are potentially influence some performance measure or quality characteristic of the process often called responses. The relationship between the response (y) and k input variables ($\xi_1,\xi_2,...,\xi_k$) could be expressed in terms of mathematical notations as follows:

$$y = f(\xi_1, \xi_2, ..., \xi_k) \tag{3}$$

where the true response function f is unknown. It is often convenient to use coded variables ($x_1,x_2,..,x_k$) instead of natural (input) variables. The response function will then be:

$$y = f(x_1, x_2, ..., x_k) \tag{4}$$

Since the form of true response function f is unknown, it must be approximated. Therefore, the successful use of RSM is critically dependent upon the choice of appropriate function to approximate f. Low-order polynomials are widely used as approximating functions. First order (linear) models are unable to capture the interaction between parameters which is a form of curvature in the true response function. Second order (quadratic) models will likely to perform well in these circumstances. In general, the quadratic model is in the form of:

$$y = \beta_0 + \sum_{j=1}^{k} \beta_j x_j + \sum_{j=1}^{k} \beta_{jj} x_j^2 + \sum_{i<j} \sum_{j=2}^{k} \beta_{ij} x_i x_j + \varepsilon \tag{5}$$

where ε is the error term in the model. The use of polynomials of higher order is also possible but infrequent. The βs are a set of unknown coefficients needed to be estimated. In order to do that, the first step is to make some observations on the system being studied. The model in Eq. (5) may now be written in matrix notations as:

$$y = X\beta + \varepsilon \tag{6}$$

where **y** is the vector of observations, X is the matrix of levels of the variables, β is the vector of unknown coefficients, and **ε** is the vector of random errors. Afterwards, method of least squares, which minimizes the sum of squares of errors, is employed to find the estimators of the coefficients ($\hat{\beta}$) through:

$$\hat{\beta} = (X'X)^{-1} X'y \tag{7}$$

The fitted model will then be written as:

$$\hat{y} = X\hat{\beta} \tag{8}$$

Finally, response surfaces or contour plots are depicted to help visualize the relationship between the response and the variables and see the influence of the parameters. As you might notice, there is a close connection between RSM and linear regression analysis.

In this chapter, RSM was employed to establish empirical relationships between four electrospinning parameters (solution concentration, spinning distance, applied voltage, and flow rate) and two responses (mean fiber diameter and standard deviation of fiber diameter). Coded variables were used to build the models. The choice of three levels for each factor in experimental design allowed us to take the advantage of quadratic models. Afterwards, the significance of terms in each model was investigated by testing hypotheses on individual coefficients and simpler yet more efficient models were obtained by eliminating statistically unimportant terms. Finally, the validity of the models was evaluated using the 15 test data. The analyses were carried out using statistical software Minitab 15.

RESULTS AND DISCUSSION

After the unknown coefficients (*βs*) were estimated by least squares method, the quadratic models for the mean fiber diameter (MFD) and standard deviation of fiber diameter (StdFD) in terms of coded variables are written as:

$$\begin{aligned} \text{MFD} = {} & 282.031 + 34.953x_1 + 5.622x_2 - 2.113x_3 + 9.013x_4 \\ & - 11.613x_1^2 - 4.304x_2^2 - 15.500x_3^2 \\ & - 0.414x_4^2 + 12.517x_1x_2 + 4.020x_1x_3 - 0.162x_1x_4 + 20.643x_2x_3 + 0.741x_2x_4 + 0.877x_3x_4 \end{aligned} \tag{9}$$

$$\begin{aligned} \text{StdFD} = {} & 36.1574 + 4.5788x_1 - 1.5536x_2 + 6.4012x_3 + 1.1531x_4 \\ & - 2.2937x_1^2 - 0.1115x_2^2 - 1.1891x_3^2 + 3.0980x_4^2 \\ & - 0.2088x_1x_2 + 1.0010x_1x_3 + 2.7978x_1x_4 + 0.1649x_2x_3 - 2.4876x_2x_4 + 1.5182x_3x_4 \end{aligned} \tag{10}$$

In the next step, a couple of very important hypothesis-testing procedures were carried out to measure the usefulness of the models presented here. First, the test for significance of the model was performed to determine whether there is a subset of variables which contributes significantly in representing the response variations. The appropriate hypotheses are:

$$\begin{aligned} & H_0 : \beta_1 = \beta_2 = \cdots = \beta_k \\ & H_1 : \beta_j \neq 0 \qquad \text{for at least one } j \end{aligned} \tag{11}$$

The *F* statistics (the result of dividing the factor mean square by the error mean square) of this test along with the *p*-values (a measure of statistical significance, the smallest level of significance for which the null hypothesis is rejected) for both models are shown in Table 9.1.

Table 9.1. Summary of the results from statistical analysis of the models.

	F	p-value	R^2	R^2_{adj}	R^2_{pred}
MFD	106.02	0.000	95.74%	94.84%	93.48%
StdFD	42.05	0.000	89.92%	87.78%	84.83%

The p-values of the models are very small (almost zero), therefore it could be concluded that the null hypothesis is rejected in both cases suggesting that there are some significant terms in each model. There are also included in Table 9.1, the values of R^2, R^2_{adj}, and R^2_{pred}. R^2 is a measure for the amount of response variation which is explained by variables and will always increase when a new term is added to the model regardless of whether the inclusion of the additional term is statistically significant or not. R^2_{adj} is the adjusted form of R^2 for the number of terms in the model; therefore it will increase only if the new terms improve the model and decreases if unnecessary terms are added. R^2_{pred} implies how well the model predicts the response for new observations, whereas R^2 and R^2_{adj} indicate how well the model fits the experimental data. The R^2 values demonstrate that 95.74% of MFD and 89.92% of StdFD are explained by the variables. The R^2_{adj} values are 94.84% and 87.78% for MFD and StdFD respectively, which account for the number of terms in the models. Both R^2 and R^2_{adj} values indicate that the models fit the data very well. The slight difference between the values of R^2 and R^2_{adj} suggests that there might be some insignificant terms in the models. Since the R^2_{pred} values are so close to the values of R^2 and R^2_{adj}, models does not appear to be over fit and have very good predictive ability.

The second testing hypothesis is evaluation of individual coefficients, which would be useful for determination of variables in the models. The hypotheses for testing of the significance of any individual coefficient are:

$$H_0 : \beta_j = 0$$
$$H_1 : \beta_j \neq 0 \qquad (12)$$

The model might be more efficient with inclusion or perhaps exclusion of one or more variables. Therefore the value of each term in the model is evaluated using this test, and then eliminating the statistically insignificant terms, more efficient models could be obtained. The results of this test for the models of MFD and StdFD are summarized in Tables 9.2 and 9.3 respectively. T statistic in these tables is a measure of the difference between an observed statistic and its hypothesized population value in units of standard error.

Table 9.2. The test on individual coefficients for the model of mean fiber diameter (MFD).

Term (coded)	**Coef**	**T**	***p*-value**
Constant	282.031	102.565	0.000
C	34.953	31.136	0.000
d	5.622	5.008	0.000

Table 9.2. *(Continued)*

Term (coded)	Coef	T	*p*-value
V	–2.113	–1.882	0.064
Q	9.013	8.028	0.000
C^2	–11.613	–5.973	0.000
d^2	–4.304	–2.214	0.030
V^2	–15.500	–7.972	0.000
Q^2	–0.414	–0.213	0.832
Cd	12.517	9.104	0.000
CV	4.020	2.924	0.005
CQ	–0.162	–0.118	0.906
dV	20.643	15.015	0.000
dQ	0.741	0.539	0.592
VQ	0.877	0.638	0.526

Table 9.3. The test on individual coefficients for the model of standard deviation of fiber diameter (StdFD).

Term (coded)	Coef	T	*p*-value
Constant	36.1574	39.381	0.000
C	4.5788	12.216	0.000
D	–1.5536	–4.145	0.000
V	6.4012	17.078	0.000
Q	1.1531	3.076	0.003
C^2	–2.2937	–3.533	0.001
d^2	–0.1115	–0.172	0.864
V^2	–1.1891	–1.832	0.072
Q^2	3.0980	4.772	0.000
Cd	–0.2088	–0.455	0.651
CV	1.0010	2.180	0.033
CQ	2.7978	6.095	0.000
dV	0.1649	0.359	0.721
dQ	–2.4876	–5.419	0.000
VQ	1.5182	3.307	0.002

As depicted, the terms related to Q^2, *CQ*, *dQ*, and *VQ* in the model of MFD and related to d^2, *Cd*, and *dV* in the model of StdFD have very high *p*-values, therefore they do not contribute significantly in representing the variation of the corresponding response. Eliminating these terms will enhance the efficiency of the models. The new models are then given by recalculating the unknown coefficients in terms of coded variables in Eqs. (13) and (14), and in terms of natural (uncoded) variables in Eqs. (15) and (16).

$$MFD = 281.755 + 34.953x_1 + 5.622x_2 - 2.113x_3 + 9.013x_4 \\ -11.613x_1^2 - 4.304x_2^2 - 15.500x_3^2 \\ +12.517x_1x_2 + 4.020x_1x_3 + 20.643x_2x_3 \quad (13)$$

$$StdFD = 36.083 + 4.579x_1 - 1.554x_2 + 6.401x_3 + 1.153x_4 \\ -2.294x_1^2 - 1.189x_3^2 + 3.098x_4^2 \\ +1.001x_1x_3 + 2.798x_1x_4 - 2.488x_2x_4 + 1.518x_3x_4 \quad (14)$$

$$MFD = 10.3345 + 48.7288C - 22.7420d + 7.9713V + 90.1250Q \\ -2.9033C^2 - 0.1722d^2 - 0.6120V^2 \\ +1.2517Cd + 0.4020CV + 0.8257dV \quad (15)$$

$$StdFD = -1.8823 + 7.5590C + 1.1818d + 1.2709V - 300.3410Q \\ -0.5734C^2 - 0.0476V^2 + 309.7999Q^2 \\ +0.1001CV + 13.9892CQ - 4.9752dQ + 3.0364VQ \quad (16)$$

The results of the test for significance as well as R^2, R^2_{adj}, and R^2_{pred} for the new models are given in Table 9.4. It is obvious that the p-values for the new models are close to zero indicating the existence of some significant terms in each model. Comparing the results of this table with Table 9.1, the F statistic increased for the new models, indicating the improvement of the models after eliminating the insignificant terms. Despite the slight decrease in R^2, the values of R^2_{adj}, and R^2_{pred} increased substantially for the new models. As it was mentioned earlier, R^2 will always increase with the number of terms in the model. Therefore, the smaller R^2 values were expected for the new models, due to the fewer terms. However, this does not necessarily suggest that the pervious models were more efficient. Looking at the tables, R^2_{adj}, which provides a more useful tool for comparing the explanatory power of models with different number of terms, increased after eliminating the unnecessary variables. Hence, the new models have the ability to better explain the experimental data. Due to higher R^2_{pred}, the new models also have higher prediction ability. In other words, eliminating the insignificant terms results in simpler models which not only present the experimental data in superior form, but also are more powerful in predicting new conditions. In a recent study , despite high reported R^2 values, researchers presented models seem to be inefficient and uncertain. Some terms in the models had very high p-values. For instance, in modeling the mean fiber diameter, p-value as high as 0.975 was calculated for cubic concentration term at spinning distance of 16 *cm*, where half of the terms had p-values more than 0.8. These results in low R^2_{pred} values which were not reported in their study and after calculating by us, they were found to be almost zero in many cases suggesting the poor prediction ability of their models.

Table 9.4. Summary of results from statistical analysis of the models after eliminating the insignificant terms.

	F	*p*-value	R^2	R^2_{adj}	R^2_{pred}
MFD	155.56	0.000	95.69%	95.08%	94.18%
StdFD	55.61	0.000	89.86%	88.25%	86.02%

The test for individual coefficients was performed again for the new models. The results of this test are summarized in Tables 9.5 and 9.6. This time, as it was anticipated, no terms had higher *p*-value than expected, which need to be eliminated. Here is another advantage of removing unimportant terms. The values of *T* statistic increased for the terms already in the models implying that their effects on the response became stronger.

Table 9.5. The test on individual coefficients for the model of MFD after eliminating the insignificant terms.

Term (coded)	Coef	T	*p*-value
Constant	281.755	118.973	0.000
C	34.953	31.884	0.000
d	5.622	5.128	0.000
V	–2.113	–1.927	0.058
Q	9.013	8.221	0.000
C^2	–11.613	–6.116	0.000
d^2	–4.304	–2.267	0.026
V^2	–15.500	–8.163	0.000
Cd	12.517	9.323	0.000
CV	4.020	2.994	0.004
dV	20.643	15.375	0.000

Table 9.6. The test on individual coefficients for the model of StdFD after eliminating the insignificant terms.

Term (coded)	Coef	T	*p*-value
Constant	36.083	45.438	0.000
C	4.579	12.456	0.000
d	–1.554	–4.226	0.000
V	6.401	17.413	0.000
Q	1.153	3.137	0.003
C^2	–2.294	–3.602	0.001
V^2	–1.189	–1.868	0.066
Q^2	3.098	4.866	0.000

Table 9.6. *(Continued)*

Term (coded)	Coef	T	*p*-value
CV	1.001	2.223	0.029
CQ	2.798	6.214	0.000
dQ	–2.488	–5.525	0.000
VQ	1.518	3.372	0.001

After developing the relationship between parameters, the test data were used to investigate the prediction ability of the models. Root mean square errors (RMSE) between the calculated responses (C_i) and real responses (R_i) were determined using Eq. (17) for experimental data as well as test data for the sake of evaluation of both MFD and StdFD models and the results are listed in Table 9.7. The models present acceptable RMSE values for test data indicating the ability of the models to generalize well the experimental data to predicting new conditions. Although the values of RMSE for the test data are slightly higher than experimental data, these small discrepancies were expected since it is almost impossible for an empirical model to express the test data as well as experimental data and higher errors are often obtained when new data are presented to the models. Hence, the results imply the acceptable prediction ability of the models.

$$\mathrm{RMSE} = \sqrt{\frac{\sum_{i=1}^{n}(C_i - R_i)^2}{n}} \qquad (17)$$

Table 9.7. RMSE values of the models for the experimental and test data.

	Experimental data	Test data
MFD	7.489	10.647
StdFD	2.493	2.890

KEYWORDS

- **Berry number (B)**
- **Response surface methodology**
- **Root mean square errors**

Chapter 10

Effects of Electric Field and Sericin Content in the Blend on the Nanofibers Uniformity

INTRODUCTION

Sericin protein is a useful biomaterial because of its unique properties. Sericin is an antibacterial and uv-resistant protein. Sericin absorbs and releases moisture easily. In this work, application of natural silk sericin to obtain antibacterial polyvinyl alcohol (PVA) nanofiber was studied. A concentrated sericin solution was dialyzed and blended with PVA solution in various Sericin/PVA blend ratios. Effects of Sericin content in the blend and electric field on the nanofibers uniformity, morphology and diameter were studied using scanning electronic microscope (SEM). Antibacterial property of electrospun mats containing sericin were also examined and reported. Results showed that the nanofibers were successfully formed up to 10% Sericin in the blends and drop formation was observed at higher sericin content. Inhibition of growing of bacteria around the PVA/sericin blend nanofiber mat was observed.

Recently, many researchers have investigated silk-based nanofibers as one of the candidate materials for biomedical applications, because it has several distinctive biological properties including good biocompatibility, good oxygen, water vapor permeability, biodegradability, and minimal inflammatory reaction . In this study nonwoven matrices of silk sericin and PVA nanofibers were prepared using electrospinning technique. Effects of electric field and sericin content in the blend on the nanofibers uniformity, morphology and diameter were studied using SEM. Antibacterial Property of Electrospun mats containing sericin were also examined and reported.

EXPERIMENTAL AND METHODS

Materials

Silk yarns made of F1 hybrid silk fibers were supplied from domestic producer Abrisham Guilan Co. Sodium carbonate (Merck), Sodium hydrogen carbonate (Merck) and other chemicals were reagent grade. PVA with molecular weight of 75,000 was obtained from Merck chemicals and distilled water was used as the solvent.

Preparation of the Spinning Solution

De-gumming of silk yarns was done in a laboratory-dyeing machine. Required amount of alkali buffer solutions at pH 10 (53.4 parts of 0.1 molar sodium carbonate + 46.6 parts of 0.1 molar sodium hydrogen carbonate) were placed in the bowls and the weighted silk yarns were immersed in the solution and sealed bowls were placed in an oil bath. Liquid to goods ratio was 40:1 and treatment temperature and time was 95°C and 30 minutes respectively. For obtaining concentrated solution of sericin, 5% aqueous acetic acid solution was added to the de-gumming wastewater and the solution pH was adjusted to 7. The solution was evaporated under reduced pressure in a rotary evaporator at 85°C. The concentrated sericin solution was dialyzed (Mw CO = 25,000) for 24 hours against distilled water. The obtained sericin solution with concentration of 20% was used for blending with 12% PVA solution. Sericin/PVA blend solutions with the weight ratio of 0/100, 2/98, 5/95, 10/90, 15/95 were prepared and subjected to the electrospinning experiments.

Electrospinning

In the electrospinning process, a high electric potential (Gamma High voltage) was applied to a droplet of PVA/Sericin solution at the tip (0.3 mm in inner diameter) of a syringe needle. The electrospun nanofibers were collected on a target plate, which was placed at a distance of 10 cm from the syringe tip. A high voltage in the range from 10 kV to 20 kV was applied to the droplet of the blend solution.

Characterization

Fiber formation and morphology of the electrospun PANI/PAN fibers were determined using an SEM Philips XL-30A (Holland). Small section of the prepared samples was placed on SEM sample holder and then coated with gold by a BAL-TEC SCD 005 sputter coater. The diameter of electrospun fibers was measured with image analyzer software (manual microstructure distance measurement). For each experiment, average fiber diameter and distribution were determined from about 100 measurements of the random fibers. Viscosity of the prepared solutions was measured using Brookfield digital viscometer (Model DVII) at 25°C.

Antibacterial Property of the electrospun mats containing Sericin was examined by Micrococoss bacteria in 37°C and bacteria growing was checked after 18 hours and was compared with PVA nanofiber mat.

RESULTS AND DISCUSSION

Figure 10.2 shows SEM micrographs of Sericin nanoparticles electrospun from pure Sericin solution. As seen in Fig. 10.2, most of Sericin particles have a round shape, while the fibrous structure is not observed. A series of experiments were carried out when the Sericin weight percent was varied from 2% to 15%. It was found that the smooth fibers were obtained when the Sericin content was up to 10% in the blend. Figures 10.1 and 10.2 show typical fiber morphology under SEM. Figure 10.3 shows viscosity of the blend solution in different blend ratio. It is clear that viscosity of the blend solutions decreases as the Sericin content increases in the blend.

Sericin/PVA blend nanofiber mat was examined for antibacterial property and for this purpose growing of the Micrococoss bacteria around the blend mat containin 5% sericin and PVA nanofiber mat has been shown in Fig. 10.4. Figure 10.5 shows that growing of the bacteria around the sericin/PVA mat has been inhibited due to the antibacterial properties of sericin whereas the bacteria has been grown around the PVA mat. This means that by blending PVA with sericin it is possible to produce antibacterial nanofiber mat for medical application.

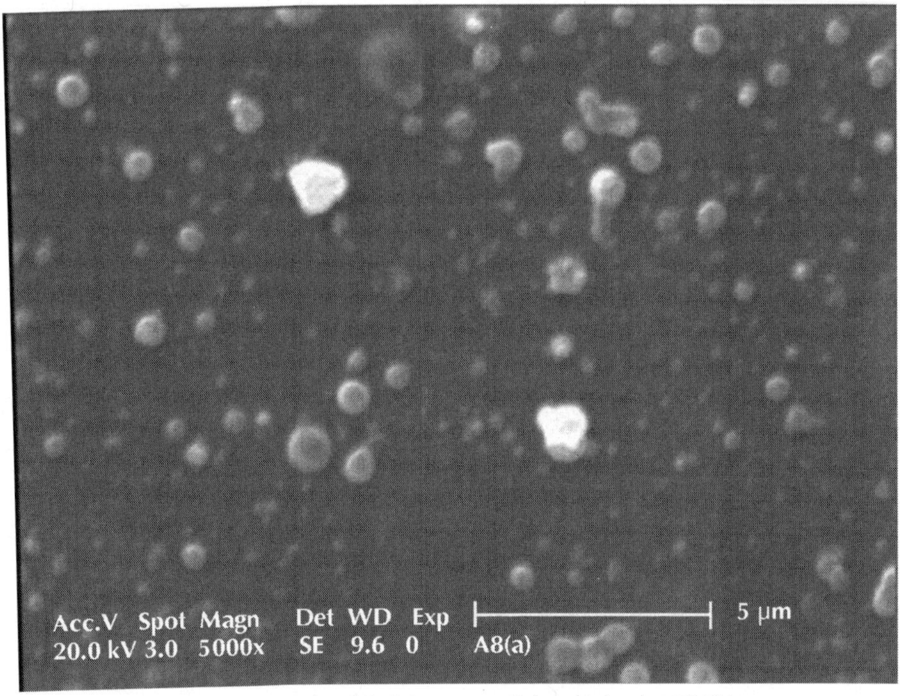

Figure 10.1. SEM photomicrographs of Sericin nanoparticles obtained at 20 KV.

Figure 10.2. SEM photomicrographs of PVA nanofibres at 20 KV.

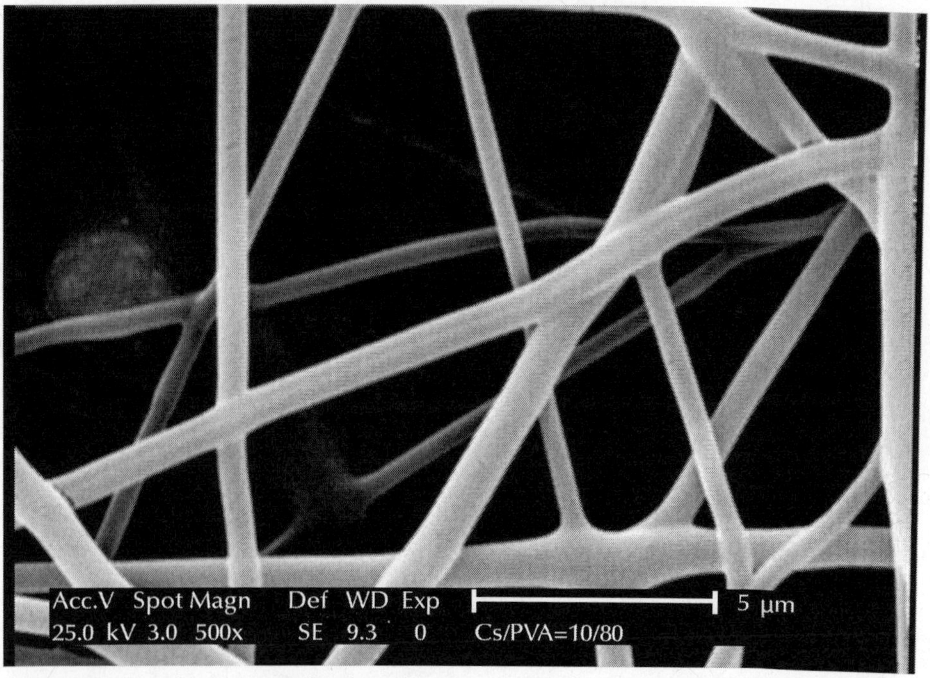

Figure 10.3. SEM photomicrographs of PVA/Sericin nanofibres at Seicin contnet of 5% and 20 KV.

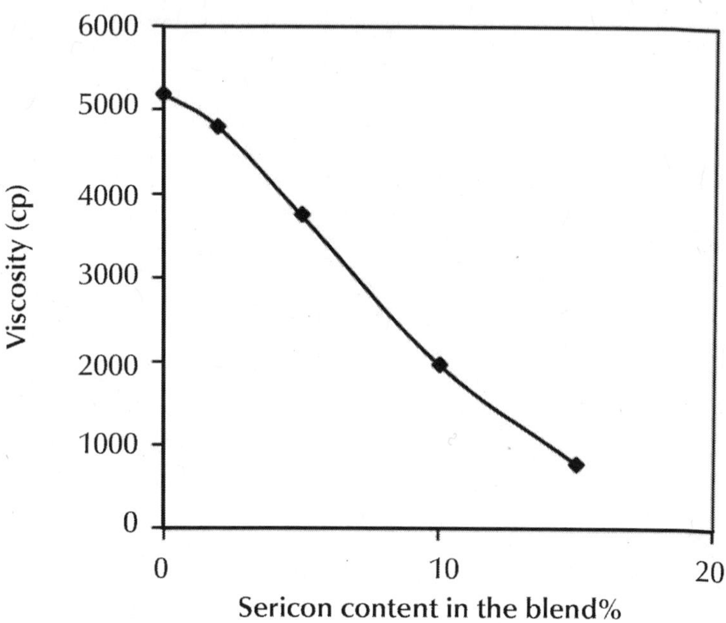

Figure 10.4. Variation of the viscosity of Sericin/PVA blend solution.

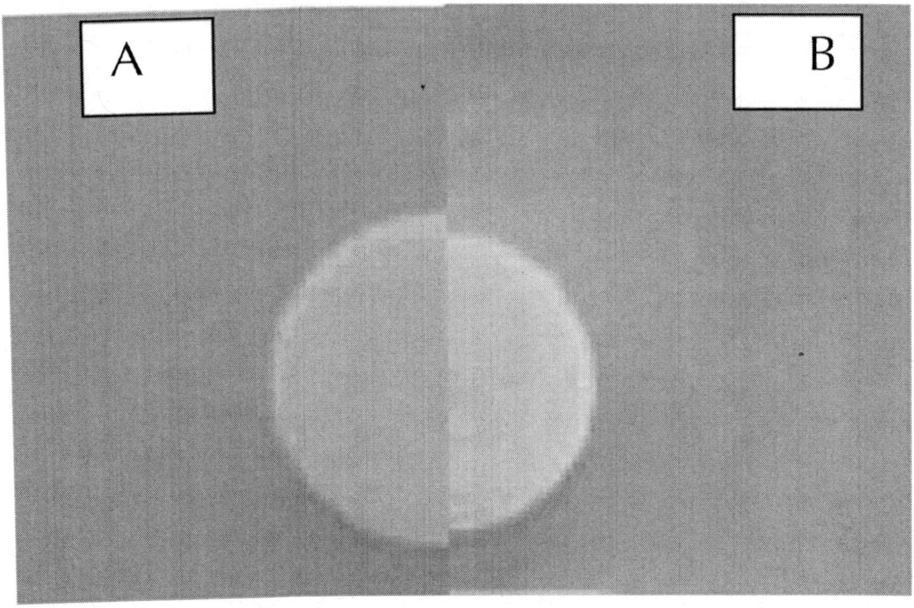

Figure 10.5. Growing of bacteria around the PVA nanofiber mat (a) and Inhibition of growing of bacteria around the PVA/sericin blend nanofibre mat at Seicin contnet of 5% (b).

KEYWORDS

- **Diameter of electrospun fibers**
- **Electrospun nanofibers**
- **Polyvinyl alcohol (PVA) mat**
- **Sericin protein**

Chapter 11

Update on Electroless plating of Fabrics with Nanoparticles

INTRODUCTION

Because of the high conductivity of copper, electroless copper plating is currently used to manufacture conductive fabrics. Electroless copper plating on fabrics has been studied by some researchers. Electroless metal plating is a non-electrolytic method of deposition from solution. The electroless deposition method uses a catalytic redox reaction between metal ions and dissolved reduction agent. This technique advantages, such as low cost, excellent conductivity, shielding effectiveness (SE), easy formation of a continuous and uniform coating on the surface of substrate with complex shapes. It can be performed at any step of the textile production, such as yarn, stock, fabric, or clothing .

Hence, chemical copper plating could be a unique process providing good potential for creation of fabrics with a metallic appearance and good handling characteristics. Revealing the performance of electroless plating of Cu-Ni-P alloy on cotton fabrics is an essential research area in textile finishing processing and for technological design. The main aim of this chapter is to explore the possibility of applying electroless plating of Cu-Ni-P alloy process onto cotton fabric. The fabrication and properties of Cu-Ni-P alloy plated cotton fabric are investigated in accordance with standard testing methods.

EXPERIMENT

Cotton fabrics (53 × 48 count/cm^2, 140 g/m^2, taffeta fabric) were used as substrate. The surface area of each specimen is 100 cm^2. The electroless copper plating process was conducted by multistep processes: pre-cleaning, sensitization, activation, electroless Cu-Ni-p alloy deposition, and post-treatment.

The fabric specimens (10 cm × 10 cm) were cleaned with non-ionic detergent (0.5g/l) and $NaHCo_3$ (0.5g/l) solution for 10 min prior to use. The samples then were rinsed in distilled water. Surface sensitization was conducted by immersion of the samples into an aqueous solution containing $SnCl_2$ and HCl. The specimens were then rinsed in deionized water and activated through immersion in an activator containing $PdCl_2$ and HCl. The substrates were then rinsed in a large volume of deionized water for 10 min to prevent contamination the plating bath. The electroless plating process carried out immediately after activation. Then all samples immersed in the electroless bath containing copper sulfate, nickel sulfate, sodium hypophosphite, sodium citrate, boric acid, and $K_4Fe\ (CN)_6$.

In the post-treatment stage, the Cu-Ni-P plated cotton fabric samples were rinsed with deionized water, ethylalcohol at home temperature for 20 min immediately after the metalizing reaction of electroless Cu-Ni-P plating. Then dried in oven at 70°C.

The weights (g) of fabric specimens with the size of 100 mm × 100 mm square before and after treatment were measured by a weight meter (HR200, AND Ltd., Japan). The percentage for the weight change of the fabric is calculated in Eq. (1).

$$I_W = \frac{W_f - W_0}{W_0} \times 100\% \tag{1}$$

where I_w is the percentage of increased weight, W_f is the final weight after treatment, W_o is the original weight.

The thickness of fabric before and after treatment was measured by a fabric thickness tester (M034A, SDL Ltd., England) with a pressure of 10 g/cm^2. The percentage of thickness increment were calculated in accordance to Eq. (2).

$$T_I = \frac{T_F - T_0}{T_0} \times 100\% \tag{2}$$

where T_I is the percentage of thickness increment, T_f is the final thickness after treatment, T_o is the original thickness.

A bending meter (003B, SDL Ltd., England) was employed to measure the degree of bending of the fabric in both warp and weft directions. The flexural rigidity of fabric samples expressed in N-cm is calculated in Eq. (3).

$$G = W \times C^3 \tag{3}$$

where G (N-cm) is the average flexural rigidity, W (N/cm^2) is the fabric mass per unit area, C (cm) is the fabric bending length.

The dimensional changes of fabrics were conducted to assess shrinkage in length for both warp and weft directions and tested with (M003A, SDL Ltd., England) accordance with standard testing method (BS EN 22313:1992). The degree of shrinkage in length expressed in percentage for both warp and weft directions were calculated in accordance to Eq. (4).

$$D_c = \frac{D_f - D_0}{D_0} \times 100 \tag{4}$$

where D_c is the average dimensional change of the treated swatch, D_o is the original dimension, D_f is the final dimension after laundering.

Tensile properties and elongation at break load were measured with standard testing method ISO 13934-1:1999 using a Micro 250 tensile tester.

Color changing under different application conditions for two standard testing methods, namely, (1) ISO 105-C06:1994 (color fastness to domestic and commercial laundering), (2) ISO 105-A02:1993 (color fastness to rubbing) were used for estimate.

Scanning electron microscope (SEM, XL30 PHILIPS) was used to characterize the surface morphology of deposits. WDX analysis (3PC, Microspec Ltd., USA) was used to exist metallic particles over surface Cu-Ni-P alloy plated cotton fabrics.

RESULTS AND DISCUSSION

Fabric Weight and Thickness

The change in weight and thickness of the untreated cotton and Cu-Ni-P alloy plated cotton fabrics are shown in Table 11.1.

Table 11.1. Weight and thickness of the untreated and Cu-Ni-P-plated cotton fabrics.

Thickness(mm)	Weight (g)	Specimen (10 cm × 10 cm)
0.4378	2.76	Untreated cotton
0.696(↑5.7 %)	3.72 (↑18.47 %)	Cu-Ni-P plated cotton

The results presented that the weight of chemically induced Cu-Ni-P-plated cotton fabric was heavier than the untreated one. The measured increased percentage of weight was 18.47%. This confirmed that Cu-Ni-P alloy had clung on the surface of cotton fabric effectively. In the case of thickness measurement, the cotton fabric exhibited a 5.7% increase after being subjected to metalization.

Fabric Bending Rigidity

Fabric bending rigidity is a fabric flexural behavior that is important for evaluating the handling of the fabric. The bending rigidity of the untreated cotton and Cu-Ni-P-plated cotton fabrics is shown in Table 11.2.

Table 11.2. Bending rigidity of the untreated and Cu-Ni-P-plated cotton fabrics.

Specimen	Bending (N-cm)	
	warp	weft
Untreated cotton	1	0.51
Cu-Ni-P plated cotton	1.17(↑11.39%)	0.66(↑30.95%)

The results proved that the chemical plating solutions had reacted with the original fabrics during the entire process of both acid sensitization and alkaline plating treatment. After electroless Cu-Ni-P alloy plating, the increase in bending rigidity level of the Cu-Ni-P-plated cotton fabrics was estimated at 11.39% in warp direction and 30.95% in weft direction respectively. The result of bending indicated that the Cu-Ni-P-plated cotton fabrics became stiffer handle than the untreated cotton fabric.

Fabric Shrinkage

The results for the fabric shrinkage of the untreated cotton and Cu-Ni-P-plated cotton fabrics are shown in Table 11.3.

Table 11.3. Dimensional change of the untreated and Cu-Ni-P-plated cotton fabrics.

Specimen	Shrinkage (%)	
	warp	**weft**
Untreated cotton	0	0
Cu-Ni-P plated cotton	–8	–13.3

The measured results demonstrated that the shrinkage level of the Cu-Ni-P-plated cotton fabric was reduced by 8% in warp direction and 13.3% in weft direction respectively.

After the Cu-Ni-P-plated, the copper particles could occupy the space between the fibers and hence more copper particles were adhered on the surface of fiber. Therefore, the surface friction in the yarns and fibers caused by the Cu-Ni-P particles could then be increased. When compared with the untreated cotton fabric, the Cu-Ni-P-plated cotton fabrics have shown a stable structure.

Fabric Tensile Strength and Elongation

The tensile strength and elongation of cotton fabrics was enhanced by the electroless Cu-Ni-P alloy plating process as shown in Table 11.4.

Table 11.4. Tensile strength and percentage of elongation at break load of the untreated and Cu-Ni-P- Fabrics plated cotton fabrics.

Specimen	Percentage of elongation (%)		Breaking load (N)	
	warp	weft	warp	weft
Untreated cotton	6.12	6.05	188.1	174.97
Cu-Ni-P plated cotton	6.98(↑12.5%)	6.52(↑7.8%)	241.5(↑28.4%)	237.3(↑35.62%)

The metalized cotton fabrics had a higher breaking load with a 28.44% increase in warp direction and a 35.62% increase in weft direction than the untreated cotton fabric. This was due to the fact that more force was required to pull the additional metal-layer coating.

The results of elongation at break were 12.5% increase in warp direction and 7.8% increase in weft direction, indicating that the Cu-Ni-P-plated fabric encountered little change when compared with the untreated cotton fabric. This confirmed that with the metalizing treatment, the specimens plated with metal particles was demonstrated a higher frictional force of fibers. In addition, the deposited metal particles which developed a linkage force to hamper the movement caused by the applied load.

Color Change Assessment

The results of evaluation of color change under different application conditions, washing, and rubbing are shown in Table 11.5.

Table 11.5. Washing and rubbing fastness of the untreated and Cu-Ni-P Fabrics plated cotton fabrics.

Specimen	Washing	Rubbing	
		Dry	Wet
Cu-Ni-P plated cotton	5	4–5	3–4

The results of the washing for the Cu-Ni-P-plated cotton fabric were grade 5 in color change. This confirms that the copper particles had good performance during washing. The result of the rubbing fastness is shown in Table 11.5. According to the test result, under dry rubbing condition, the degree of staining was recorded to be grade 4–5, and the wet rubbing fastness showed grade 3–4 in color change. These results showed that the dry rubbing fastness had a lower color change in comparison with the wet crocking fastness. In view of the overall results, the rubbing fastness of the Cu-Ni-P-plated cotton fabric was relatively good when compared with the commercial standard.

Surface Morphology

Scanning electron microscopy (SEM) of the untreated and Cu-Ni-P-plated cotton fabric is shown in Fig. 11.1 with magnification of 650x. Microscopic evidents of copper coated fabrics shows the formation of evenness copper particles on fabric surface and structure.

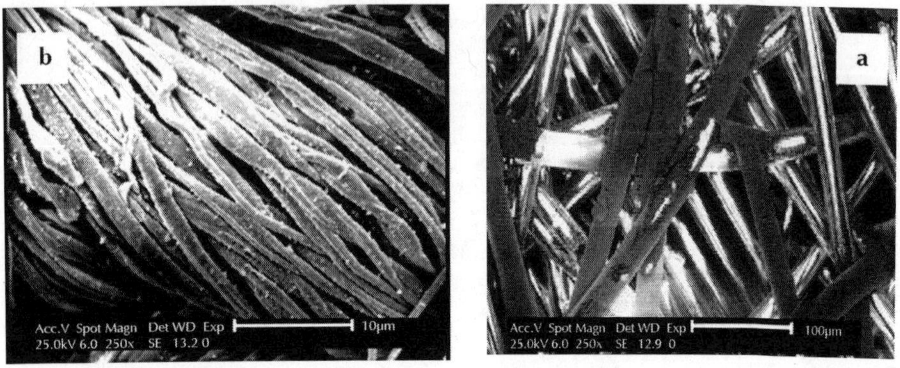

Figure 11.1. SEM photographs of the (a) untreated cotton fabric (b) Cu-Ni-P plated cotton fabric.

Figure 11.2 shows the SEM and WDX analysis copper plated surfaces of cotton fiber. It was observed that the cotton fibers surface was covered by Cu-Ni-P alloy particles composing of an evenly distributed mass. In addition, WDX analysis indicated that the deposits became more compact, uniform, and smoother also exist nanoparticle metal over surface copper-coated fabrics. These results indicated that the effect of chemical copper plating was sufficient and effective to change the micro-structure of the cotton fabric.

Figure11.2. (a) SEM photograph of the Cu-Ni-P plated cotton fabric (b) WDX analysis of the Cu-Ni-P plated cotton fabric.

Shielding Effectiveness

Electromagnetic shielding means that the energy of electromagnetic radiation is attenuated by reflection or absorption of an electromagnetic shielding material, which is one of the effective methods to realize electromagnetic compatibility. The unit of EMISE is given in decibels (dB). The EMISE value was calculated from the ratio of the incident to transmitted power of the electromagnetic wave in the following Eq. 5.

$$SE = 10\log\left|\frac{P_1}{P_2}\right| = 20\left|\frac{E_1}{E_2}\right| \tag{5}$$

where P_1 (E_1) and P_2 (E_2) are the incident power (incident electric field) and the transmitted power (transmitted electric field), respectively. Figure 11.3 indicates the SE of

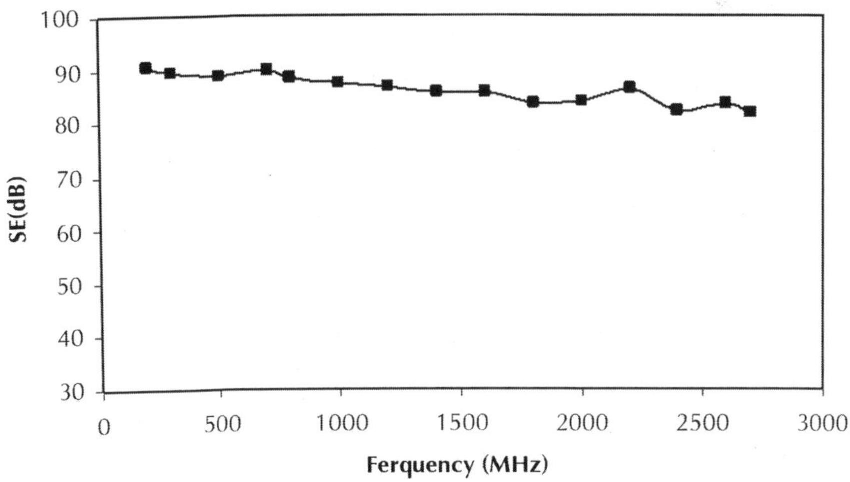

Figure 11.3. The shielding effectiveness of copper-coated cotton fabrics.

the uncoated and copper-coated fabrics with 1 ppm $K_4Fe\ (CN)_6$. As a result, the SE of cotton fabrics was almost zero at the frequencies 50 MHz to 1.5 GHz. However, SE of copper-coated cotton fabric was above 90 dB and the tendency of SE kept similarity at the frequencies 50 MHz to 1.5 GHz. The copper-coated cotton fabric has a practical usage for many EMI shielding application requirements.

KEYWORDS

- **Bending meter**
- **Electroless deposition method**
- **Electroless metal plating**
- **Electromagnetic shielding**
- **Fabric bending rigidity**
- **Scanning electron microscope**

Chapter 12

Update on Lamination of Nanofibers

INTRODUCTION

Polymeric nanofibers can be made using the electrospinning process, has already been described in the literature. Electrospinning (Fig. 12.1) uses a high electric field to draw a polymer solution from tip of a capillary toward a collector. A voltage is applied to the polymer solution, which causes a jet of the solution to be drawn toward a grounded collector. The fine jets dry to form polymeric fibers, which can be collected as a web .

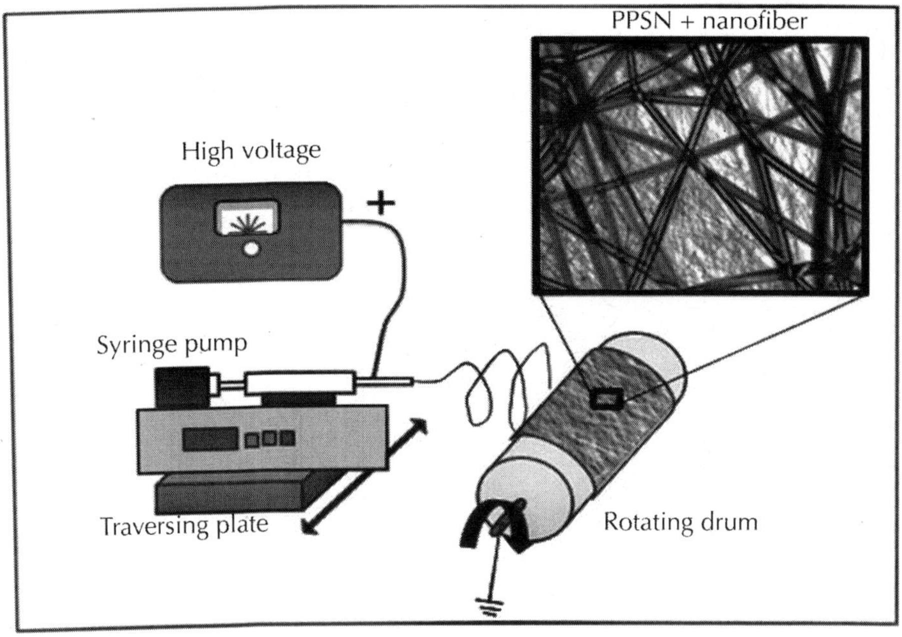

Figure 12.1. Electrospinning setup.

In the nonwoven industry one of the fastest growing segments is in filtration applications. Traditionally wet-laid, melt blown, and spun nonwoven articles, containing micron size fibers are most popular for these applications because of the low cost, easy process ability, and good filtration efficiency. Their applications in filtration can be divided into two major areas: air filtration and liquid filtration (Fig. 12.2).

Figure 12.2. Multilayer fabric components.

Another type of electrospinning equipment (Fig. 12.2) also used a variable high voltage power supply from Gamma High Voltage Research (USA). The applied voltage can be varied from 1–30 kV. A 5 ml syringe was used and positive potential was applied to the polymer blend solution by attaching the electrode directly to the outside of the hypodermic needle with internal diameter of 0.3 mm. The collector screen was a 20 × 20 cm aluminum foil, which was placed 10 cm horizontally from the tip of the needle. The electrode of opposite polarity was attached to the collector. A metering syringe pump from New Era pump systems Inc. (USA) was used. It was responsible for supplying polymer solution with a constant rate of 20 μl/min.

Electrospinning was done in a temperature-controlled chamber and temperature of electrospinning environment was adjusted on variable temperatures. Schematic diagram of the electrospinning apparatus is shown in Fig. 12.3.

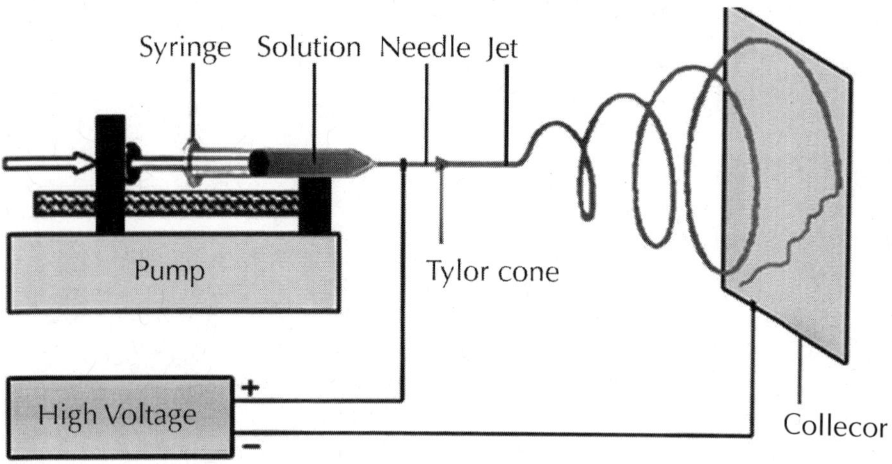

Figure 12.3. Schematic diagram of general type of electrospinning apparatus.

Elecrospinning is a process that produces continuous polymer fibers with diameter in the sub-micron range. In the electrospinnig process the electric body force act on element of charged fluid. Electrospinning has emerged as a specialized processing technique for the formation of sub-micron fibers (typically between 100 nm and 1 μm in diameter), with high specific surface areas. Due to their high specific surface area, high porosity, and small pore size, the unique fibers have been suggested for excellent candidate for use in filtration .

Air and water are the bulk transportation medium for transmission of particulate contaminants. The contaminants during air filtration are complex mixture of particles. The most of them are usually smaller than 1000 μm in diameter chemical and biological aerosols are frequently in range of 1–10 μm. The particulate matters may carry some gaseous contaminants. In water filtration removal of particulate and biological contaminants is an important step. Nowadays, the filtration industry is looking for energy efficient high performance filters for filtration of particles smaller than 0.3 μm and adsorbed toxic gases .

Nanofibrous media have low basis weight, high permeability, and small pore size that make them appropriate for a wide range of filtration applications. In addition, nanofiber membrane offers unique properties like high specific surface area (ranging from 1 to 35m^2/g depending on the diameter of fibers), good interconnectivity of pores and potential to incorporate active chemistry or functionality on nanoscale. In our study, to prepare the filters, a flow rate 1 μl/h for solution was selected and the fibers were collected on an aluminum-covered rotating drum (with speed 9 m/min) which was previously covered with a polypropylene spun-bond nonwoven (PPSN) substrate of 28 cm × 28 cm dimensions; 0.19 mm thickness; 25 g/m2 weight; 824 $cm^3/s/cm^2$ air permeability; and 140°C melting point (Fig. 12.2).

Structure characteristics of nanofiberous filtering media such as layer thickness ,fiber diameter, nanofiber orientation, representative pore size ,porosity dictate the filter properties, and quality. Clearly, the properties of a nanofiberous media will depend on its structural characteristics, as well as the nature of the component fibers. Thus, it is desirable to understand and determine these characteristics. In this work tried to identify the orientation distribution function (ODF) of nanofibers in nanofilter, the fiber thickness distribution, and porosity of nanofiberous media by using image processing algorithms.

Effect of Systematic Parameters on Electrospun Nanofibers

It has been found that morphology such as fiber diameter and its uniformity of the electrospun nanofibers are dependent on many processing parameters. These parameters can be divided into three main groups: (1) solution properties, (2) processing conditions, and (3) ambient conditions. Each of the parameters has been found to affect the morphology of the electrospun fibers.

Solution Properties

Parameters such as viscosity of solution, solution concentration, molecular weight of solution, electrical conductivity, elasticity, and surface tension, have important effect on morphology of nanofibers.

Viscosity

The viscosity range of a different nanofiber solution which is spinnable is different. One of the most significant parameters influencing the fiber diameter is the solution viscosity. A higher viscosity results in a large fiber diameter. Beads and beaded fibers are less likely to be formed for the more viscous solutions. The diameter of the beads become bigger and the average distance between beads on the fibers longer as the viscosity increases.

Solution Concentration

In electrospinning process, for fiber formation to occur, a minimum solution concentration is required. As the solution concentration increase, a mixture of beads and fibers is obtained. The shape of the beads changes from spherical to spindle-like when the solution concentration varies from low to high levels. It should be noted that the fiber diameter increases with increasing solution concentration because the higher viscosity resistance. Nevertheless, at higher concentration, viscoelastic force which usually resists rapid changes in fiber shape may result in uniform fiber formation. However, it is impossible to electrospin if the solution concentration or the corresponding viscosity become too high due to the difficulty in liquid jet formation.

Molecular Weight

Molecular weight also has a significant effect on the rheological and electrical properties such as viscosity, surface tension, conductivity, and dielectric strength. It has been observed that too low molecular weight solution tend to form beads rather than fibers and high molecular weight nanofiber solution give fibers with larger average diameter.

Surface Tension

The surface tension of a liquid is often defined as the force acting at right angles to any line of unit length on the liquid surface. By reducing surface tension of a nanofiber solution, fibers could be obtained without beads. The surface tension seems more likely to be a function of solvent compositions, but is negligibly dependent on the solution concentration. Different solvents may contribute different surface tensions. However, not necessarily a lower surface tension of a solvent will always be more suitable for electrospinning. Generally, surface tension determines the upper and lower boundaries of electrospinning window if all other variables are held constant. The formation of droplets, bead, and fibers can be driven by the surface tension of solution and lower surface tension of the spinning solution helps electrospinning to occur at lower electric field.

Solution Conductivity

There is a significant drop in the diameter of the electrospun nanofibers when the electrical conductivity of the solution increases. Beads may also be observed due to low conductivity of the solution, which results in insufficient elongation of a jet by electrical force to produce uniform fiber. In general, electrospun nanofibers with the smallest fiber diameter can be obtained with the highest electrical conductivity. This interprets that the drop in the size of the fibers is due to the increased electrical conductivity.

Applied Voltage

In the case of electrospinning, the electric current due to the ionic conduction of charge in the nanofiber solution is usually assumed small enough to be negligible. The only mechanism of charge transport is the flow of solution from the tip to the target. Thus, an increase in the electrospinning current generally reflects an increase in the mass flow rate from the capillary tip to the grounded target when all other variables (conductivity, dielectric constant, and flow rate of solution to the capillary tip) are held constant. Increasing the applied voltage (i.e., increasing the electric field strength) will increase the electrostatic repulsive force on the fluid jet which favors the thinner fiber formation. On the other hand, the solution will be removed from the capillary tip more quickly as jet is ejected from Taylor cone. This results in the increase of the fiber diameter.

Feed Rate

The morphological structure can be slightly changed by changing the solution flow rate. When the flow rate exceeded a critical value, the delivery rate of the solution jet to the capillary tip exceeds the rate at which the solution was removed from the tip by the electric forces. This shift in the mass-balance resulted in sustained but unstable jet and fibers with big beads formation.

In the first part of this study, the production of electrospun nanofibers investigated. In another part, a different case study presented to show how nanofibers can be laminated for application in filter media.

EXPERIMENTS: CASE 1—PRODUCTION OF NANOFIBERS

Preparation of Regenerated SF Solution

Raw silk fibers (B. mori cocoons were obtained from domestic producer, Abrisham Guilan Co., Iran) were degummed with 2 gr/L Na_2CO_3 solution and 10 gr/L anionic detergent at 100°C for 1 h and then rinsed with warm distilled water. Degummed silk fibroin (SF) was dissolved in a ternary solvent system of $CaCl_2/CH_3CH_2OH/H_2O$ (1:2:8 in molar ratio) at 70°C for 6 h. After dialysis with cellulose tubular membrane (Bialysis Tubing D9527 Sigma) in H_2O for 3 days, the SF solution was filtered and lyophilized to obtain the regenerated SF sponges.

Preparation of the Spinning Solution

SF solutions were prepared by dissolving the regenerated SF sponges in 98% formic acid for 30 min. Concentrations of SF solutions for electrospinning was in the range from 8% to 14% by weight.

Electrospinning

In the electrospinning process, a high electric potential (Gamma High voltage) was applied to a droplet of SF solution at the tip (0.35 mm inner diameter) of a syringe needle, The electrospun nanofibers were collected on a target plate which was placed at a distance of 10 cm from the syringe tip. The syringe tip and the target plate were enclosed in a chamber for adjusting and controlling the temperature. Schematic diagram of the electrospinning apparatus is shown in Fig. 12.2. The processing temperature was adjusted at 25, 50, and 75°C. A high voltage in the range from 10 kV to 20 kV was applied to the droplet of SF solution.

Characterization

Optical microscope (Nikon Microphot-FXA) was used to investigate the macroscopic morphology of electrospun SF fibers. For better resolving power, morphology, surface texture, and dimensions of the gold-sputtered electrospun nanofibers were determined using a Philips XL-30 scanning electron microscope. A measurement of about 100 random fibers was used to determine average fiber diameter and their distribution.

EXPERIMENT: CASE 2—PRODUCTION OF LAMINATED COMPOSITES

Polyacrylonitrile (PAN) of 70,000 g/mol molecular weight from Polyacryl Co. (Isfehan, Iran) has been used with Dimethylformamide (DMF) from Merck, to form a polymer solution 12% w/w after stirring for 5 h and staying overnight under room temperature. The yellow and ripen solution was inserted into a plastic syringe with a stainless steel nozzle 0.4 mm in inner diameter and then it was placed in a metering pump from World Precision Instruments (Florida, USA). Next, this set installed on a plate which it could traverse to left–right along drum (Fig. 12.1). The flow rate 1 μl/h for solution was selected and the fibers were collected on an aluminum-covered rotating drum (with speed 9 m/min) which was previously covered with a PPSN substrate of 28×28cm dimensions; 0.19 mm thickness; 25 g/m2 weight; 824 $cm^3/s/cm^2$ air permeability; and 140°C melting point. The distance between the nozzle and the drum was 7 cm and an electric voltage of approximately 11 kV was applied between them. Electrospinning process was carried out for 8 h at room temperature to reach approximately web thickness 3.82 g/m^2. Then nanofiber webs were laminated into cotton weft–warp fabric with a thickness 0.24 mm and density of 25 × 25 (warp–weft) per centimeter to form a multilayer fabric (Fig. 12.2). Laminating was performed at temperatures 85, 110, 120, 140, and 160°C for 1 min under a pressure of 9 gf/cm^2.

Air permeability of multilayer fabric before and after lamination was tested by TEXTEST FX3300 instrument (Zürich, Switzerland). Also, in order to consider of nanofiber morphology after hot-pressing, another laminating was performed by a non-stick sheet made of Teflon (0.25 mm thickness) instead one of the fabrics (fabric/pp web/nanofiber web/pp web/non-stick sheet). Finally, after removing of Teflon sheet, the nanofiber layer side was observed under an optical microscope (MICROPHOT-FXA, Nikon, Japan) connected to a digital camera.

RESULTS AND DISCUSSION

Effect of Silk Concentration

One of the most important quantities related with electrospun nanofibers is their diameter. Since nanofibers are resulted from evaporation of polymer jets, the fiber diameters will depend on the jet sizes and the solution concentration. It has been reported that during the traveling of a polymer jet from the syringe tip to the collector, the primary jet may be split into different sizes multiple jets, resulting in different fiber diameters. When no splitting is involved in electrospinning, one of the most important parameters influencing the fiber diameter is concentration of regenerated silk solution. The jet with a low concentration breaks into droplets readily and a mixture of fibers, bead fibers, and droplets as a result of low viscosity is generated. These fibers have an irregular morphology with large variation in size, on the other hand jet with high concentration do not break up but traveled to the grounded target and tend to facilitate the formation of fibers without beads and droplets. In this case, fibers became more uniform with regular morphology.

At first, a series of experiments were carried out when the silk concentration was varied from 8 to 14% at the 15 kV constant electric field and 25°C constant temperature. Below the silk concentration of 8%, as well as at low electric filed in the case of 8% solution, droplets were formed instead of fibers. Figure 12.2 shows morphology of the obtained fibers from 8% silk solution at 20 kV. The obtained fibers are not uniform. The average fiber diameter is 72 nm and a narrow distribution of fiber diameters is observed. It was found that continues nanofibers were formed above silk concentration of 8% regardless of the applied electric field and electrospinning condition. In the electrospinning of silk fibroin, when the silk concentration is more than 10%, thin and rod like fibers with diameters range from 60 to 450 nm were obtained.

There is a significant increase in mean fiber diameter (MFD) with the increasing of the silk concentration, which shows the important role of silk concentration in fiber formation during electrospinning process. Concentration of the polymer solution reflects the number of entanglements of polymer chains in the solution, thus solution viscosity. Experimental observations in electrospinning confirm that for forming fibers, a minimum polymer concentration is required. Below this critical concentration, application of electric field to a polymer solution results electrospraying and formation of droplets to the instability of the ejected jet. As the polymer concentration increased, a mixture of beads and fibers is formed. Further increase in concentration results in formation of continuous fibers as reported in this chapter. It seems that the critical concentration of the silk solution in formic acid for the formation of continuous silk fibers is 10%.

Experimental results in electrospinning showed that with increasing the temperature of electrospinning process, concentration of polymer solution has the same effect on fibers diameter at 25°C.

There is a significant increase in MFD with increasing of the silk concentration, which shows the important role of silk concentration in fiber formation during electrospinning process. It is well known that the viscosity of polymer solutions is proportional to concentration and polymer molecular weight. For concentrated polymer solution, concentration of the polymer solution reflects the number of entanglements of polymer chains, thus have considerable effects on the solution viscosity. At fixed polymer molecular weight, the higher polymer concentration resulting higher solution viscosity. The jet from low viscosity liquids breaks up into droplets more readily and few fibers are formed, while at high viscosity, electrospinning is prohibit because of the instability flow causes by the high cohesiveness of the solution. Experimental observations in electrospinning confirm that for fiber formation to occur, a minimum polymer concentration is required. Below this critical concentration, application of electric field to a polymer solution results electrospraying and formation of droplets to the instability of the ejected jet. As the polymer concentration increased, a mixture of beads and fibers is formed. Further increase in concentration results in formation of continuous fibers as reported in this chapter. It seems that the critical concentration of the silk solution in formic acid for the formation of continuous silk fibers is 10% when the applied electric field was in the range of 10–20 kV.

Effect of Electric Field

It was already reported that the effect of the applied electrospinning voltage is much lower than effect of the solution concentration on the diameter of electrospun fibers. In order to study the effect of the electric field, silk solution with the concentration of 10%, 12%, and 14% were electrospun at 10, 15, and 20 kV at 25°C. At a high solution concentration, effect of applied voltage is nearly significant. It is suggested that, at this temperature, higher applied voltage causes multiple jets formation, which would provide decrees fiber diameter.

As the results of this finding it seems that electric field shows different effects on the nanofibers morphology. This effect depends on the polymer solution concentration and electrospinning conditions.

Effect of Electrospinning Temperature

One of the most important quantities related with electrospun nanofibers is their diameter. Since nanofibers are resulted from evaporation of polymer jets, the fiber diameters will depend on the jet sizes. The elongation of the jet and the evaporation of the solvent both change the shape and the charge per unit area carried by the jet. After the skin is formed, the solvent inside the jet escapes and the atmospheric pressure tends to collapse the tube like jet. The circular cross section becomes elliptical and then flat, forming a ribbon-like structure. In this work, we believe that ribbon-like structure in the electrospinning of SF at higher temperature thought to be related with skin formation at the jets. With increasing the electrospinning temperature, solvent evaporation rate increases, which results in the formation of skin at the jet surface. Non-uniform lateral stresses around the fiber due to the uneven evaporation of solvent and/or striking the target make the nanofibers with circular cross-section to collapse into ribbon shape.

Bending of the electrospun ribbons were observed on the SEM micrographs as a result of the electrically driven bending instability or forces that occurred when the ribbon was stopped on the collector. Another problem that may be occurring in the electrospinning of SF at high temperature is the branching of jets. With increasing the temperature of electrospinning process, the balance between the surface tension and electrical forces can shift so that the shape of a jet becomes unstable. Such an unstable jet can reduce its local charge per unit surface area by ejecting a smaller jet from the surface of the primary jet or by splitting apart into two smaller jets. Branched jets, resulting from the ejection of the smaller jet on the surface of the primary jet were observed in electrospun fibers of SF. The axes of the cones from which the secondary jets originated were at an angle near 90° with respect to the axis of the primary jet.

In order to study the effect of electrospinning temperature on the morphology and texture of electrospun silk nanofibers, 12% silk solution was electrospun at various temperatures of 25, 50, and 75°C. Results are shown in Fig. 12.4. Interestingly, the electrospinning of silk solution showed flat fiber morphology at 50 and 75°C, whereas circular structure was observed at 25°C. At 25°C, the nanofibers with a rounded cross section and a smooth surface were collected on the target. Their diameter showed a size range of approximately 100–300 nm with 180 nm being the most frequently occurring. They are within the same range of reported size for electrospun silk nanofibers. With increasing the electrospinning temperature to 50°C, the morphology of the fibers was slightly changed from circular cross section to ribbon-like fibers. Fiber diameter was also increased to a range of approximately 20–320 nm with 180 nm the most occurring frequency. At 75°C, the morphology of the fibers was completely changed to ribbon-like structure. Furthermore, fibers dimensions were increased significantly to the range of 500–4100 nm with 1100 nm the most occurring frequency. The results are shown in Fig. 12.4.

EXPERIMENTAL DESIGN

Response surface methodology (RSM) is a collection of mathematical and statistical techniques for empirical model building (Appendix). By careful design of experiments, the objective is to optimize a response (output variable) which is influenced by several independent variables (input variables). An experiment is a series of tests, called runs, in which changes are made in the input variables in order to identify the reasons for changes in the output response.

In order to optimize and predict the morphology and average fiber diameter of electrospun silk, design of experiment was employed in the present work. Morphology of fibers and distribution of fiber diameter of silk precursor were investigated varying concentration, temperature, and applied voltage. A more systematic understanding of these process conditions was obtained and a quantitative basis for the relationships between average fiber diameter and electrospinning parameters was established using RSM, which will provide some basis for the preparation of silk nanofibers.

A central composite design was employed to fit a second-order model for three variables. Silk concentration (X_1), applied voltage (X_2), and temperature (X_3) were three independent variables (factors) considered in the preparation of silk nanofibers, while the fibers diameter were dependent variables (response). The actual and corresponding coded values of three factors (X_1, X_2, and X_3) are given in Table 12.1. The following second-order model in X_1, X_2, and X_3 was fitted using the data in Table 12.1:

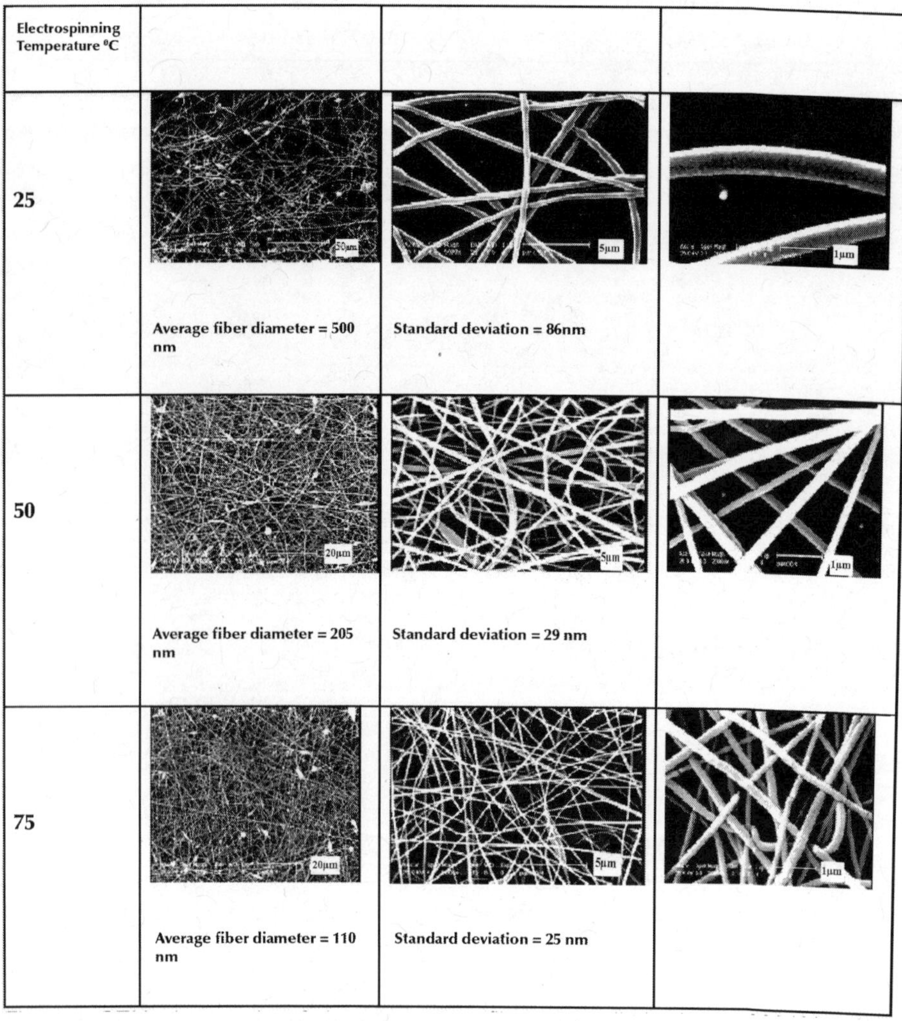

Figure 12.4. SEM micrographs of electrospun nanofibers at applied voltage of 20 kV and PANI content of 20% with a constant spinning distance of 10 cm.

$$Y = \beta_0 + \beta_1 x_1 + \beta_2 x_2 + \beta_3 x_3 + \beta_1 x_1^2 + \beta_2 x_2^2 + \beta_3 x_3^2 + \beta_2 x_1 x_2 + \beta_3 x_1 x_3 + \beta_3 x_2 x_3 + \varepsilon \quad (1)$$

Table 12.1. Central composite design.

X_i	Independent variables	Coded values		
		–1	0	1
X_1	silk concentration (%)	10	12	14
X_2	applied voltage (kV)	10	15	20
X_3	temperature (°C)	25	50	75

The MINITAB and MATHLAB programs were used for analysis of this second-order model and for response surface plots (MINITAB 11, MATHLAB 7).

By regression analysis, values for coefficients for parameters and *p*-values (a measure of the statistical significance) are calculated. When *p*-value is less than 0.05, the factor has significant impact on the average fiber diameter. If *p*-value is greater than 0.05, the factor has no significant impact on average fiber diameter. And R^2_{adj} (represents the proportion of the total variability that has been explained by the regression model) for regression models were obtained (Table 12.2). The fitted second-order equation for average fiber diameter can be considered by:

$$Y = 391 + 311 X_1 - 164 X_2 + 57 X_3 - 162 X_1^{2} + 69 X_2^{2} + 391 X_3^{2} - 159 X_1X_2 + 315 X_1X_3 - 144 X_2X_3 \quad (2)$$

where Y = Average fiber diameter

Table 12.2. Regression Analysis for the three factors (concentration, applied voltage, temperature) and coefficients of the model in coded unit*.

Variables	**Constant**		***p*-value**
	β_0	391.3	0.008
x_1	β_1	310.98	0.00
x_2	β_2	–164.0	0.015
x_3	β_3	57.03	0.00
x^2_1	β_{11}	161.8	0.143
x^2_2	β_{22}	68.8	0.516
x^2_3	β_{33}	390.9	0.002
x_1x_2	β_{12}	–158.77	0.048
x_1x_3	β_{13}	314.59	0.001
x_2x_3	β_{23}	–144.41	0.069
F	*p*-value	R^2	R^2 (adj)
18.84	0.00	0.907	0.858

*Model: $Y=\beta_0+\beta_1x_1+\beta_2x_2+\beta_3x_3+\beta_{11}x^2_1+\beta_{22}x^2_2+\beta_{332}x^2_3+\beta_{12}x_1x2+\beta_{13}x_1x\,3+\beta_{13}x_2x3$ where "y" is average fiber diameter.

From the *p*-values listed in Table 12.2, it is obvious that *p*-value of term X_2 is greater than *p*-values for terms X_1 and X_3. And other *p*-values for terms related to applied voltage such as, X_2^2, X_1X_2, and X_2X_3 are much greater than significance level of 0.05. That is to say, applied voltage has no much significant impact on average fiber diameter and the interactions between concentration and applied voltage, temperature, and applied voltage are not significant, either. But *p*-values for term related to X_3 and X_1 are less than 0.05. Therefore, temperature and concentration have significant impact on average fiber diameter. Furthermore, R^2_{adj} is 0.858, That is to say, this model explains 86% of the variability in new data (see Figs. 12.5 and 12.6).

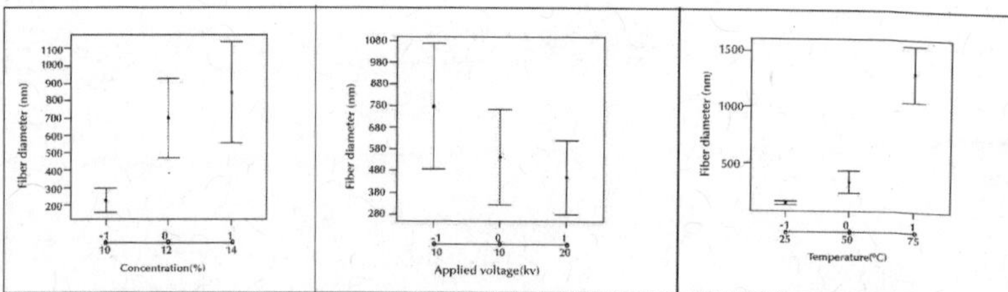

Relation between average fiber diameter and solutions concentration	Relation between average fiber diameter and applied voltage	Relation between average fiber diameter and posses temperature

Figure 12.5. Effect of electrospinning parameters on nanofibers diameter.

Laminating temperature (°C)	Taken images at 100 magnification	Observations
85		* Inconspicuous melting of PPSN
110	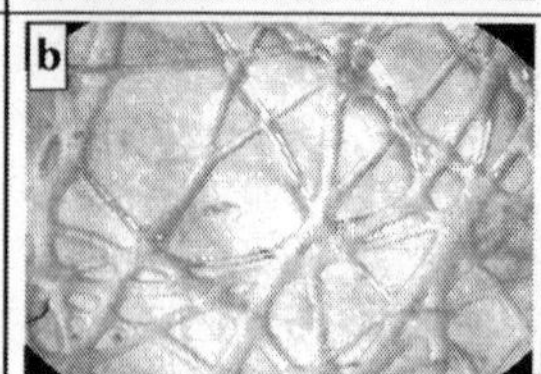	* Incomplete melting of PPSN and creation of linkage with nanofiber web
120	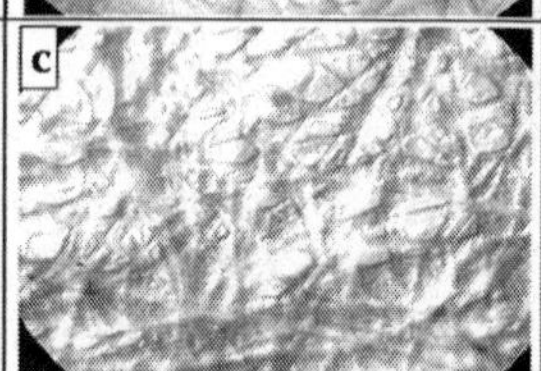	* Incomplete melting of PPSN and creation of linkage with nanofiber web *Spreading of melt over nanofiber surface because of pressing effect *Increasing junction area
140		* Complete melting of PPSN with phase change *Penetration of melt into nanofiber web
160 (>140)	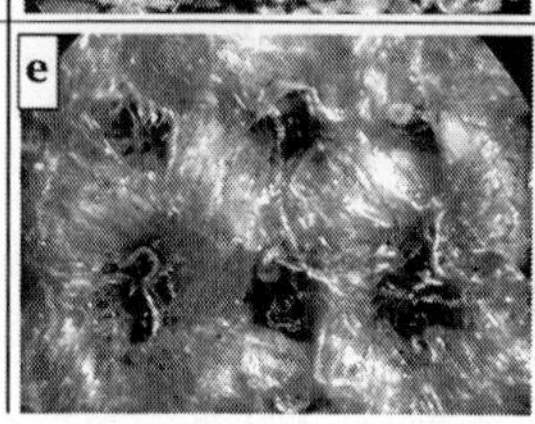	* Whole penetration of melt into nanofiber structure and web missing

Figure 12.6. The optical microscope images of nanofiber web after lamination at various temperatures to be used as filter media.

APPENDIX

Variables which potentially can alter the electrospinning process (Fig. A-1) are large. Hence, investigating all of them in the framework of one single research would almost be impossible. However, some of these parameters can be held constant during experimentation. For instance, performing the experiments in a controlled environmental condition, which is concerned in this study, the ambient parameters (i.e., temperature, air pressure, and humidity) are kept unchanged. Solution viscosity is affected by polymer molecular weight, solution concentration, and temperature. For a particular polymer (constant molecular weight) at a fixed temperature, solution concentration would be the only factor influencing the viscosity. In this circumstance, the effect of viscosity could be determined by the solution concentration. Therefore, there would be no need for viscosity to be considered as a separate parameter.

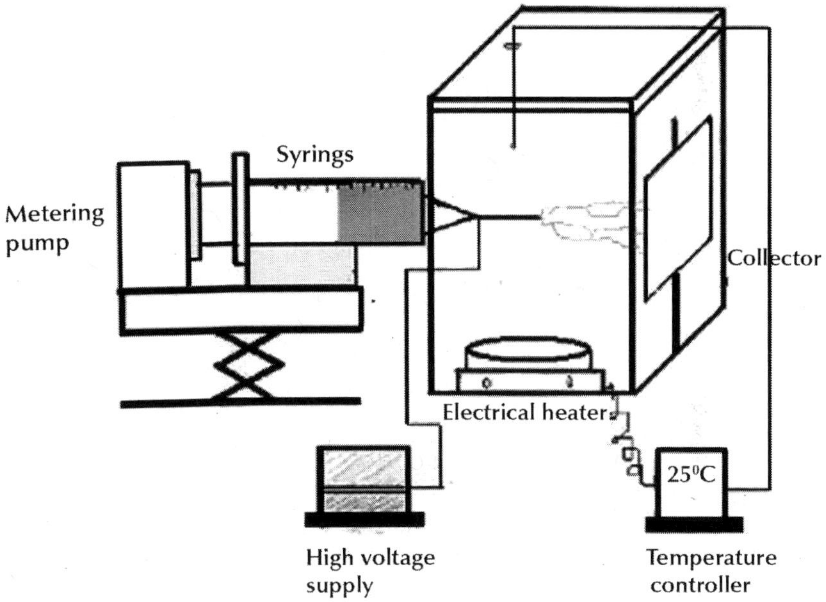

Figure A-1. A typical image of electrospinning process.

In this regard, solution concentration (C), spinning distance (d), applied voltage (V), and volume flow rate (Q) were selected to be the most influential parameters. The next step is to choose the ranges over which these factors are varied. Process knowledge, which is a combination of practical experience and theoretical understanding, is required to fulfill this step. The aim is here to find an appropriate range for each parameter where dry, bead-free, stable, and continuous fibers without breaking up to droplets are obtained. This goal could be achieved by conducting a set of preliminary experiments while having the previous works in mind along with utilizing the reported relationships.

The relationship between intrinsic viscosity ($[\eta]$) and molecular weight (M) is given by the well-known Mark–Houwink–Sakurada equation as follows:

$$[\eta] = KM^a \tag{A-1}$$

where K and a are constants for a particular polymer-solvent pair at a given temperature. Polymer chain entanglements in a solution can be expressed in terms of Berry number (B), which is a dimensionless parameter and defined as the product of intrinsic viscosity and polymer concentration ($B=[\eta]C$). For each molecular weight, there is a lower critical concentration at which the polymer solution cannot be electrospun.

As for determining the appropriate range of applied voltage, referring to previous works, it was observed that the changes of voltage lay between 5 *kV* and 25 *kV* depending on experimental conditions; voltages above 25 *kV* were rarely used. Afterwards, a series of experiments were carried out to obtain the desired voltage domain. At $V<10$ *kV*, the voltage was too low to spin fibers and $10\ kV \leq V < 15\ kV$ resulted in formation of fibers and droplets; in addition, electrospinning was impeded at high concentrations. In this regard, $15\ kV \leq V \leq 25\ kV$ was selected to be the desired domain for applied voltage.

The use of 5 *cm*–20 *cm* for spinning distance was reported in the literature. Short distances are suitable for highly evaporative solvents. whereas it results in wet coagulated fibers for nonvolatile solvents due to insufficient evaporation time. Afterwards, this was proved by experimental observations and $10\ cm \leq d \leq 20\ cm$ was considered as the effective range for spinning distance.

Few researchers have addressed the effect of volume flow rate. Therefore in this case, the attention was focused on experimental observations. At $Q<0.2\ ml/h$, in most cases especially at high polymer concentrations, the fiber formation was hindered due to insufficient supply of solution to the tip of the syringe needle. Whereas, excessive feed of solution at $Q>0.4\ ml/h$ incurred formation of droplets along with fibers. As a result, $0.2\ ml/h \leq Q \leq 0.4\ ml/h$ was chosen as the favorable range of flow rate in this study.

Consider a process in which several factors affect a response of the system. In this case, a conventional strategy of experimentation, which is extensively used in practice, is the *one-factor-at-a-time* approach. The major disadvantage of this approach is its failure to consider any possible interaction between the factors, say the failure of one factor to produce the same effect on the response at different levels of another factor. For instance, suppose that two factors A and B affect a response. At one level of A, increasing B causes the response to increase, while at the other level of A, the effect of B totally reverses and the response decreases with increasing B. As interactions exist between electrospinning parameters, this approach may not be an appropriate choice for the case of the present work. The correct strategy to deal with several factors is to use a full factorial design. In this method, factors are all varied together; therefore all possible combinations of the levels of the factors are investigated. This approach is very efficient, makes the most use of the experimental data and takes into account the interactions between factors.

It is trivial that in order to draw a line at least two points and for a quadratic curve at least three points are required. Hence, three levels were selected for each parameter in this study so that it would be possible to use quadratic models. These levels were chosen equally spaced. A full factorial experimental design with four factors (solution concentration, spinning distance, applied voltage, and flow rate) each at three levels (3^4 design) were employed resulting in 81 treatment combinations. This design is shown in Fig. A-2.

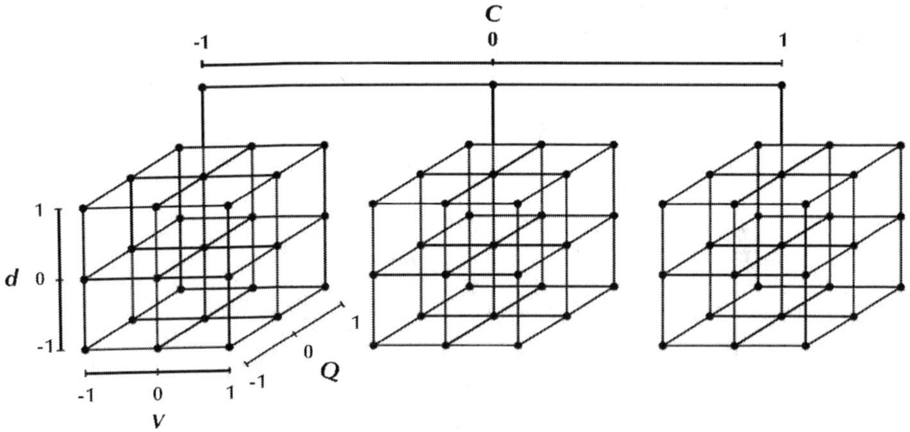

Figure A-2. 3^4 full factorial experimental design used in this study.

Coded variables –1, 0, and 1 are corresponding to low, intermediate, and high levels of each factor respectively. The coded variables (x_j) were calculated using Eq. (A-2) from natural variables (ξ_i). The indices 1–4 represent solution concentration, spinning distance, applied voltage, and flow rate respectively. In addition to experimental data, 15 treatments inside the design space were selected as test data and used for evaluation of the models. The natural and coded variables for experimental data (numbers 1–81), as well as test data (numbers 82–96) are listed in Table 8 in Appendix.

$$x_j = \frac{\xi_j - [\xi_{hj} + \xi_{lj}]/2}{[\xi_{hj} - \xi_{lj}]/2} \tag{A-2}$$

The mechanism of some scientific phenomena has been well understood and models depicting the physical behavior of the system have been drawn in the form of mathematical relationships. However, there are numerous processes at the moment which have not been sufficiently understood to permit the theoretical approach. RSM is a combination of mathematical and statistical techniques useful for empirical modeling and analysis of such systems. The application of RSM is in situations where several input variables are potentially influence some performance measure or quality characteristic of the process—often called responses. The relationship between the response (y) and k input variables ($\xi_1,\xi_2,...,\xi_k$) could be expressed in terms of mathematical notations as follows:

$$y = f(\xi_1, \xi_2, ..., \xi_k) \tag{A-3}$$

where the true response function f is unknown. It is often convenient to use coded variables (x_1, x_2,..,x_k) instead of natural (input) variables. The response function will then be:

$$y = f(x_1, x_2, ..., x_k) \tag{A-4}$$

Since the form of true response function f is unknown, it must be approximated. Therefore, the successful use of RSM is critically dependent upon the choice of appropriate function to approximate f. Low-order polynomials are widely used as approximating functions. First-order (linear) models are unable to capture the interaction between parameters which is a form of curvature in the true response function. Second-order (quadratic) models will be likely to perform well in these circumstances. In general, the quadratic model is in the form of:

$$y = \beta_0 + \sum_{j=1}^{k} \beta_j x_j + \sum_{j=1}^{k} \beta_{jj} x_j^2 + \sum_{i<j} \sum_{j=2}^{k} \beta_{ij} x_i x_j + \varepsilon \tag{A-5}$$

where ε is the error term in the model. The use of polynomials of higher order is also possible but infrequent. The βs are a set of unknown coefficients needed to be estimated. In order to do that, the first step is to make some observations on the system being studied. The model in Eq. (A-5) may now be written in matrix notations as:

$$y = X\beta + \varepsilon \tag{A-6}$$

where $\mathbf{y}$ is the vector of observations, X is the matrix of levels of the variables, β is the vector of unknown coefficients, and $\boldsymbol{\varepsilon}$ is the vector of random errors. Afterwards, method of least squares, which minimizes the sum of squares of errors, is employed to find the estimators of the coefficients ($\hat{\beta}$) through:

$$\hat{\beta} = (X'X)^{-1} X'y \tag{A-7}$$

The fitted model will then be written as:

$$\hat{y} = X\hat{\beta} \tag{A-8}$$

Finally, response surfaces or contour plots are depicted to help visualize the relationship between the response and the variables and see the influence of the parameters. As you might notice, there is a close connection between RSM and linear regression analysis.

After the unknown coefficients (βs) were estimated by least squares method, the quadratic models for the MFD and standard deviation of fiber diameter (StdFD) in terms of coded variables are written as:

$$\begin{aligned}\text{MFD} = {} & 282.031 + 34.953x_1 + 5.622x_2 - 2.113x_3 + 9.013x_4 \\ & -11.613x_1^2 - 4.304x_2^2 - 15.500x_3^2 \\ & -0.414x_4^2 + 12.517x_1x_2 + 4.020x_1x_3 - 0.162x_1x_4 + 20.643x_2x_3 + 0.741x_2x_4 + 0.877x_3x_4\end{aligned} \quad \text{(A-9)}$$

$$\begin{aligned}\text{StdFD} = {} & 36.1574 + 4.5788x_1 - 1.5536x_2 + 6.4012x_3 + 1.1531x_4 \\ & -2.2937x_1^2 - 0.1115x_2^2 - 1.1891x_3^2 + 3.0980x_4^2 \\ & -0.2088x_1x_2 + 1.0010x_1x_3 + 2.7978x_1x_4 + 0.1649x_2x_3 - 2.4876x_2x_4 + 1.5182x_3x_4\end{aligned} \quad \text{(A-10)}$$

In the next step, a couple of very important hypothesis-testing procedures were carried out to measure the usefulness of the models presented here. First, the test for significance of the model was performed to determine whether there is a subset of variables which contributes significantly in representing the response variations. The appropriate hypotheses are:

$$\begin{aligned} & H_0 : \beta_1 = \beta_2 = \cdots = \beta_k \\ & H_1 : \beta_j \neq 0 \quad \text{for at least one } j \end{aligned} \quad \text{(A-11)}$$

The F statistics (the result of dividing the factor mean square by the error mean square) of this test along with the p-values (a measure of statistical significance, the smallest level of significance for which the null hypothesis is rejected) for both models are shown in Table A-1.

Table A-1. Summary of the results from statistical analysis of the models.

	F	p-value	R^2	R^2_{adj}	R^2_{pred}
MFD	106.02	0.000	95.74%	94.84%	93.48%
StdFD	42.05	0.000	89.92%	87.78%	84.83%

The p-values of the models are very small (almost zero), therefore it could be concluded that the null hypothesis is rejected in both cases suggesting that there are some significant terms in each model. There are also included in Table A-1, the values of R^2, R^2_{adj}, and R^2_{pred}. R^2 is a measure for the amount of response variation which is explained by variables and will always increase when a new term is added to the model regardless of whether the inclusion of the additional term is statistically significant or not. R^2_{adj} is the adjusted form of R^2 for the number of terms in the model; therefore it will increase only if the new terms improve the model and decreases if unnecessary terms are added. R^2_{pred} implies how well the model predicts the response for new observations, whereas R^2 and R^2_{adj} indicate how well the model fits the experimental data. The R^2 values demonstrate that 95.74% of MFD and 89.92% of StdFD are explained by the variables. The R^2_{adj} values are 94.84% and 87.78% for MFD and StdFD respectively, which account for the number of terms in the models. Both R^2 and R^2_{adj} values indicate that the models fit the data very well. The slight difference between the values of R^2 and R^2_{adj} suggests that there might be some insignificant terms in the models. Since the R^2_{pred} values are so close to the values of R^2 and R^2_{adj}, models does not appear to be overfit and have very good predictive ability.

The second testing hypothesis is evaluation of individual coefficients, which would be useful for determination of variables in the models. The hypotheses for testing of the significance of any individual coefficient are:

$$H_0 : \beta_j = 0$$
$$H_1 : \beta_j \neq 0 \qquad \text{(A-12)}$$

The model might be more efficient with inclusion or perhaps exclusion of one or more variables. Therefore, the value of each term in the model is evaluated using this test, and then eliminating the statistically insignificant terms, more efficient models could be obtained. The results of this test for the models of MFD and StdFD are summarized in Tables A-2 and A-3 respectively. *T* statistic in these tables is a measure of the difference between an observed statistic and its hypothesized population value in units of standard error.

Table A-2. The test on individual coefficients for the model of MFD.

Term (coded)	Coeff.	T	*p*-value
Constant	282.031	102.565	0.000
C	34.953	31.136	0.000
D	5.622	5.008	0.000
V	–2.113	–1.882	0.064
Q	9.013	8.028	0.000
C^2	–11.613	–5.973	0.000
d^2	–4.304	–2.214	0.030
V^2	–15.500	–7.972	0.000
Q^2	–0.414	–0.213	0.832
Cd	12.517	9.104	0.000
CV	4.020	2.924	0.005
CQ	–0.162	–0.118	0.906
dV	20.643	15.015	0.000
dQ	0.741	0.539	0.592
VQ	0.877	0.638	0.526

Table A-3. The test on individual coefficients for the model of StdFD.

Term (coded)	Coeff.	T	*p*-value
Constant	36.1574	39.381	0.000
C	4.5788	12.216	0.000
D	–1.5536	–4.145	0.000
V	6.4012	17.078	0.000
Q	1.1531	3.076	0.003
C^2	–2.2937	–3.533	0.001
d^2	–0.1115	–0.172	0.864
V^2	–1.1891	–1.832	0.072
Q^2	3.0980	4.772	0.000
Cd	–0.2088	–0.455	0.651
CV	1.0010	2.180	0.033
CQ	2.7978	6.095	0.000
dV	0.1649	0.359	0.721
dQ	–2.4876	–5.419	0.000
VQ	1.5182	3.307	0.002

As depicted, the terms related to Q^2, *CQ*, *dQ*, and *VQ* in the model of MFD and related to d^2, *Cd*, and *dV* in the model of StdFD have very high *p*-values, therefore they do not contribute significantly in representing the variation of the corresponding response. Eliminating these terms will enhance the efficiency of the models. The new models are then given by recalculating the unknown coefficients in terms of coded variables in Eqs. (A-13) and (A-14), and in terms of natural (uncoded) variables in Eqs. (A-15) and (A-16).

$$\begin{aligned} MFD = {} & 281.755 + 34.953x_1 + 5.622x_2 - 2.113x_3 + 9.013x_4 \\ & -11.613x_1^2 - 4.304x_2^2 - 15.500x_3^2 \\ & +12.517x_1x_2 + 4.020x_1x_3 + 20.643x_2x_3 \end{aligned} \tag{A-13}$$

$$\begin{aligned} StdFD = {} & 36.083 + 4.579x_1 - 1.554x_2 + 6.401x_3 + 1.153x_4 \\ & -2.294x_1^2 - 1.189x_3^2 + 3.098x_4^2 \\ & +1.001x_1x_3 + 2.798x_1x_4 - 2.488x_2x_4 + 1.518x_3x_4 \end{aligned} \tag{A-14}$$

$$\begin{aligned} MFD = {} & 10.3345 + 48.7288C - 22.7420d + 7.9713V + 90.1250Q \\ & -2.9033C^2 - 0.1722d^2 - 0.6120V^2 \\ & +1.2517Cd + 0.4020CV + 0.8257dV \end{aligned} \tag{A-15}$$

$$StdFD = -1.8823 + 7.5590C + 1.1818d + 1.2709V - 300.3410Q \\ -0.5734C^2 - 0.0476V^2 + 309.7999Q^2 \\ +0.1001CV + 13.9892CQ - 4.9752dQ + 3.0364VQ \tag{A-16}$$

The results of the test for significance, as well as R^2, R^2_{adj}, and R^2_{pred} for the new models are given in Table A-4. It is obvious that the *p*-values for the new models are close to zero indicating the existence of some significant terms in each model. Comparing the results of Table A-4 *with* Table A-1, the *F* statistic increased for the new models, indicating the improvement of the models after eliminating the insignificant terms. Despite the slight decrease in R^2, the values of R^2_{adj}, and R^2_{pred} increased substantially for the new models. As it was mentioned earlier in the chapter, R^2 will always increase with the number of terms in the model. Therefore, the smaller R^2 values were expected for the new models, due to the fewer terms. However, this does not necessarily suggest that the previous models were more efficient. Looking at the tables, R^2_{adj}, which provides a more useful tool for comparing the explanatory power of models with different number of terms, increased after eliminating the unnecessary variables. Hence, the new models have the ability to better explain the experimental data. Due to higher R^2_{pred}, the new models also have higher prediction ability.

Table A-4. Summary of the results from statistical analysis of the models after eliminating the insignificant terms.

	F	*p*-value	R^2	R^2_{adj}	R^2_{pred}
MFD	155.56	0.000	95.69%	95.08%	94.18%
StdFD	55.61	0.000	89.86%	88.25%	86.02%

The test for individual coefficients was performed again for the new models. The results of this test are summarized in Tables A-5 and A-6. This time, as it was anticipated, no terms had higher *p*-value than expected, which need to be eliminated. Here is another advantage of removing unimportant terms. The values of *T* statistic increased for the terms already in the models implying that their effects on the response became stronger.

Table A-5. The test on individual coefficients for the model of mean fiber diameter (MFD) after eliminating the insignificant terms.

Term (coded)	Coeff.	T	*p*-value
Constant	281.755	118.973	0.000
C	34.953	31.884	0.000
d	5.622	5.128	0.000
V	–2.113	–1.927	0.058
Q	9.013	8.221	0.000

Table A-5. *(Continued)*

Term (coded)	Coeff.	T	*p*-value
C^2	–11.613	–6.116	0.000
d^2	–4.304	–2.267	0.026
V^2	–15.500	–8.163	0.000
Cd	12.517	9.323	0.000
CV	4.020	2.994	0.004
dV	20.643	15.375	0.000

Table A-6. The test on individual coefficients for the model of StdFD after eliminating the insignificant terms.

Term (coded)	Coeff.	T	*p*-value
Constant	36.083	45.438	0.000
C	4.579	12.456	0.000
d	–1.554	–4.226	0.000
V	6.401	17.413	0.000
Q	1.153	3.137	0.003
C^2	–2.294	–3.602	0.001
V^2	–1.189	–1.868	0.066
Q^2	3.098	4.866	0.000
CV	1.001	2.223	0.029
CQ	2.798	6.214	0.000
dQ	–2.488	–5.525	0.000
VQ	1.518	3.372	0.001

After developing the relationship between parameters, the test data were used to investigate the prediction ability of the models. Root mean square errors (RMSE) between the calculated responses (C_i) and real responses (R_i) were determined using Eq. (A-17) for experimental data, as well as test data for the sake of evaluation of both MFD and StdFD models.

$$\mathrm{RMSE} = \sqrt{\frac{\sum_{i=1}^{n}(C_i - R_i)^2}{n}} \tag{A-17}$$

KEYWORDS

- **Berry number**
- **Degummed silk fibroin**
- **Electrospinning**
- **MINITAB and MATHLAB programs**
- **Raw silk fibers**
- **Response surface methodology**

Chapter 13

Update on New Class of Nonwovens

INTRODUCTION

Nowadays, there are different types of protective clothing that some of these are disposable and non-disposable. The simplest and most preliminary of this equipment is made from rubber or plastic that is completely impervious to hazardous substances. Unfortunately, these materials are also impervious to air and water vapor, and thus retain body heat, exposing their wearer to heat stress which can build quite rapidly to a dangerous level. Another approach to protective clothing is incorporating activated carbon into multilayer fabric in order to absorb toxic vapors from environment and prevent penetration to the skin. The use of activated carbon is considered only a short term solution because it loses its effectiveness upon exposure to sweat and moisture. The use of semi-permeable membranes as a constituent of the protective material is another approach. In this way, reactive chemical decontaminants encapsulates in microparticles or fills in microporous hollow fibers and then coats onto fabric. The microparticle or fiber walls are permeable to toxic vapors, but impermeable to decontaminants, so that the toxic agents diffuse selectively into them and neutralize. Generally, a negative relationship always exists between thermal comfort and protection performance for currently available protective clothing. Thus, there still exists a very real demand for improved protective clothing that can offer acceptable levels of impermeability to highly toxic pollutions of low molecular weight, while minimizing wearer discomfort and heat stress.

Electrospinning provides an ultrathin membrane-like web of extremely fine fibers with very small pore size and high porosity, which makes them excellent candidates for use in filtration, membrane, and possibly protective clothing applications. Preliminary investigations have indicated that the using of nanofiber web in protective clothing structure could present minimal impedance to air permeability and extremely efficiency in trapping dust and aerosol particles. Meanwhile, it is found that the electrospun webs of nylon 6, 6, polybenzimidazole, polyacrylonitrile, and polyurethane provided good aerosol particle protection, without a considerable change in moisture vapor transport or breathability of the system. While nanofiber webs suggest exciting characteristics, it has been reported that they have limited mechanical properties. In order to provide suitable mechanical properties for use as cloth, nanofiber webs must be laminated via an adhesive into a fabric system. This system could protect ultrathin nanofiber web versus mechanical stresses over an extended period of time.

The adhesives could be as melt adhesive or solvent-based adhesive. When a melt adhesive is used, the hot-press laminating carried out at temperatures above the softening or melting point of adhesive. If a solvent-based adhesive is used, laminating process could perform at room temperature. In addition, the solvent-based adhesive is generally environmentally unfriendly, more expensive and usually flammable, whereas the hot-melt adhesive is environmentally friendly, inexpensive requires less heat, and so is now more preferred. However, without disclosure of laminating details, the hot-press method is more suitable for nanofiber web lamination. In this method, laminating temperature is one of the most important parameters. Incorrect selection of this parameter may lead to change or damage nanofiber web. Thus, it is necessary to find out a laminating temperature which has the least effect on the nanofiber web.

It has been found that morphology such as fiber diameter and its uniformity of the electrospun polymer fibers are dependent on many processing parameters. These parameters can be divided into three groups as shown in Table 13.1. Under certain condition, not only uniform fibers but also beads-like formed fibers can be produced by electrospinning. Although the parameters of the electrospinning process have been well analyzed in each of polymers these information has been inadequate enough to support the electrospinning of ultra-fine nanometer scale polymer fibers. A more systematic parametric study is hence required to investigate.

Table 13.1. Tensile strength test results of the multilayer fabrics.

Multilayer Fabric	*Warp direction*			
	Breaking Load, N		*Breaking Elongation, mm*	
	Mean value	**CV, %**	**Mean value**	**CV, %**
Without Nanofiber web	174.427	6.2	5.02	7.5
With Nanofiber web	189.211	4.6	5.11	6

The purpose of this study is to consider the influence of laminating temperature on nanofiber/laminate properties. Multilayer fabrics were made by electrospinnig polyacrylonitrile nanofibers onto nonwoven substrate and incorporating into fabric system via hot-press method at different temperatures.

EXPERIMENT

Electrospining and Laminating Process

Polyacrylonitrile (PAN) of 70,000 g/mol molecular weight from an industrial sector has been used with Dimethylformamide (DMF) from Merck, to form a polymer solution 12% w/w after stirring for 5h and exposing for 24h at ambient temperature. The yellow and ripen solution was inserted into a plastic syringe with a stainless steel nozzle 0.4 mm in inner diameter and then it was placed in a metering pump from World Precision Instruments (Florida, USA). Next, this set installed on a plate which it could traverse to left-right along drum (Fig. 13.1). The flow rate 1 μl/h for solution was selected and the fibers were collected on an aluminum-covered rotating drum (with speed 9 m/min) which was previously covered with a polypropylene spun-bond nonwoven (PPSN) substrate of 28cm×28cm dimensions; 0.19 mm thickness; 25 g/m^2 weight; 824 cm^3/s/cm^2 air permeability and 140°C melting point. The distance between the nozzle and the drum was 7 cm and an electric voltage of approximately 11kV was applied between them. Electrospinning process was carried out for 8h at room temperature to reach approximately web thickness 3.82 g/m². Then nanofiber webs were laminated into cotton weft-warp fabric with a thickness 0.24 mm and density of 25×25 (warp-weft) per centimeter to form a multilayer fabric (Fig. 13.2). Laminating was performed at temperatures 85, 110, 120, 140, 150°C for 1 min under a pressure of 9 gf/cm².

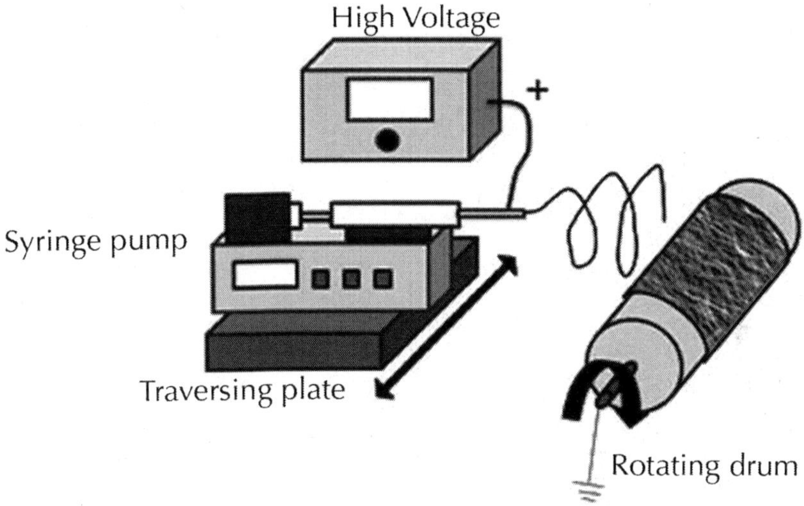

Figure 13.1. Electrospinning setup.

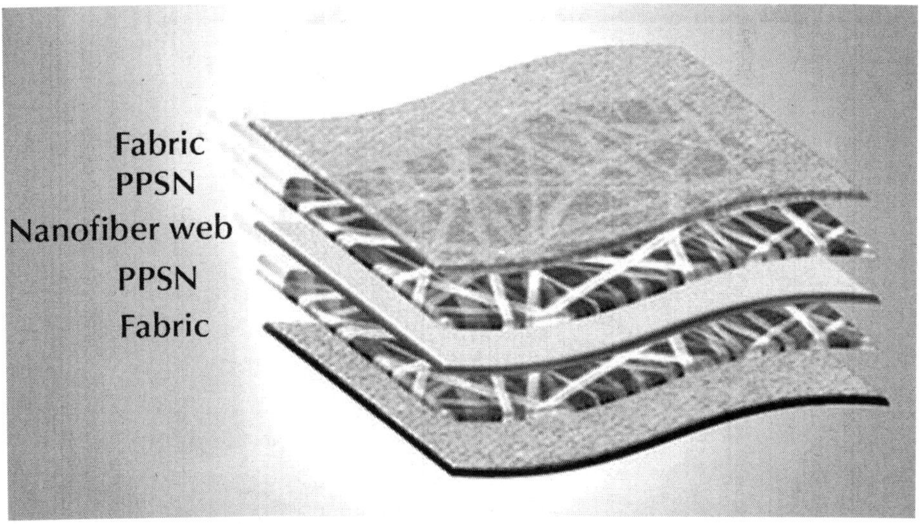

Figure 13.2. Multilayer fabric components.

Nanofiber Web Morphology

In order to consider of nanofiber web morphology after hot-pressing, another laminating was performed by a non-stick sheet made of Teflon (0.25 mm thickness) instead one of the fabrics (fabric/pp web/nanofiber web/pp web/non-stick sheet). Finally, after removing of Teflon sheet, the nanofiber layer side was observed under an optical microscope (MICROPHOT-FXA, Nikon, Japan) connected to a digital camera.

Measurement of Air Permeability

Air permeability of multilayer fabric after lamination was tested by TEXTEST FX3300 instrument (Zürich, Switzerland). It was tested 5 pieces of each sample under air pressure 125 pa at ambient condition (16°C, 70% RH) and obtained average air permeability.

RESULTS AND DISCUSSION

PPSN was selected as melt adhesive layer for hot-press laminating (Fig. 13.2). This process was performed under different temperatures to find an optimum condition. Figure 13.3 presents the optical microscope images of nanofiber web after lamination. It is obvious that by increasing of laminating temperature to melting point (samples a-c) the adhesive layer gradually melts and spreads on web surface. But, when melting point selected as laminating temperature (sample *d*) the nanofiber web begin to damage. In this case, the adhesive layer completely melted and penetrated into nanofiber web and occupied its pores. This procedure intensified by increasing of laminating

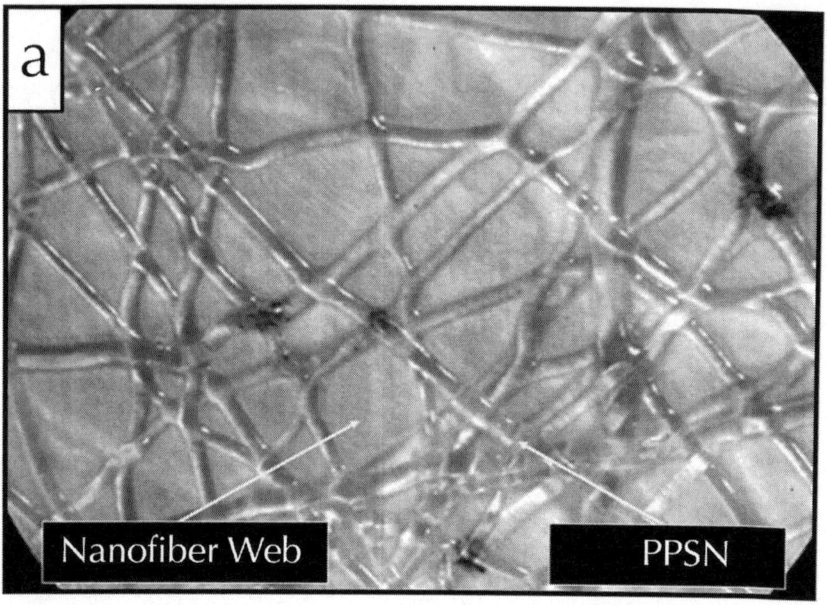

Figure 13.3. (a) The optical microscope images of nanofiber web after laminating at 85°C (at 100 magnification).

temperature above melting point. As shown in Fig. 13.1 (sample *e*), perfect absorption of adhesive by nanofiber web creates a transparent film which leads to appear fabric structure.

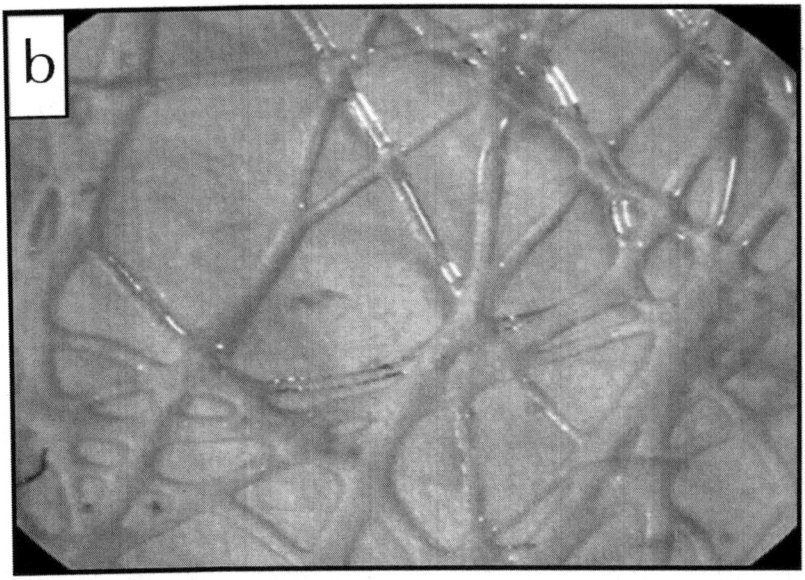

Figure 13.3. (b) The optical microscope images of nanofiber web after laminating at 110°C (at 100 magnification).

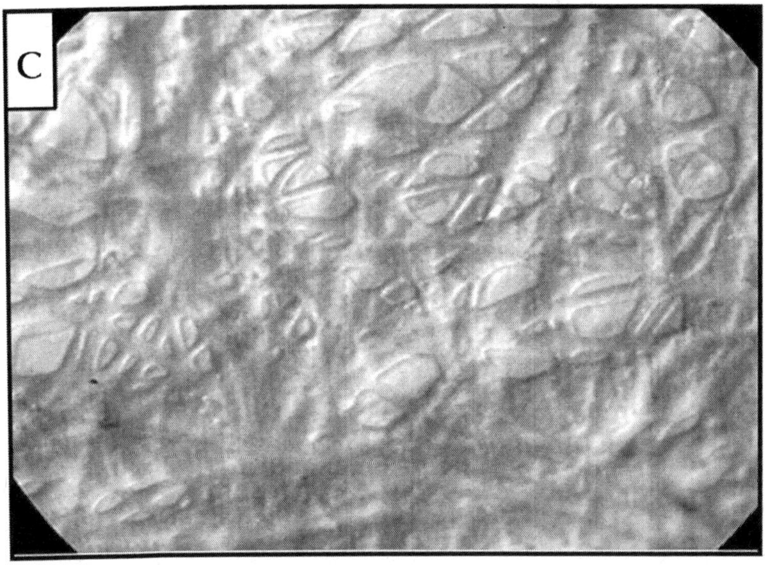

Figure 13.3. (c) The optical microscope images of nanofiber web after laminating at 120°C (at 100 magnification).

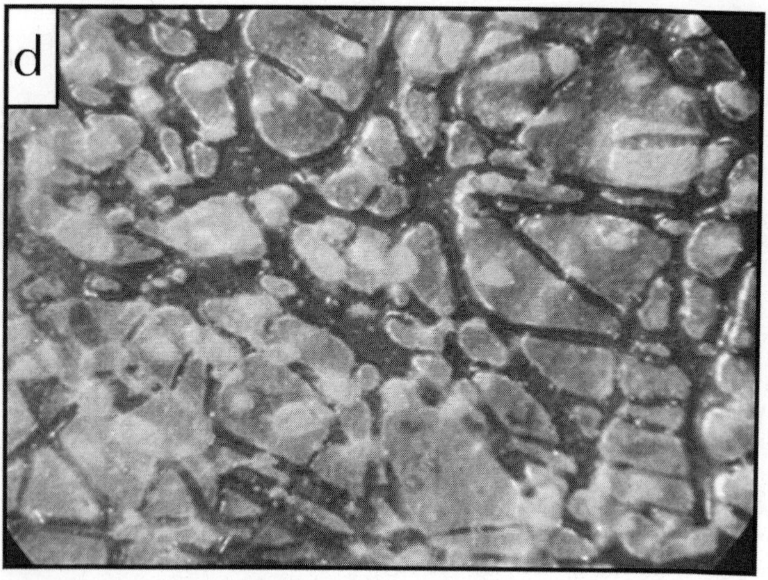

Fig.ure 13.3. (d) The optical microscope images of nanofiber web after laminating at 140°C (at 100 magnification).

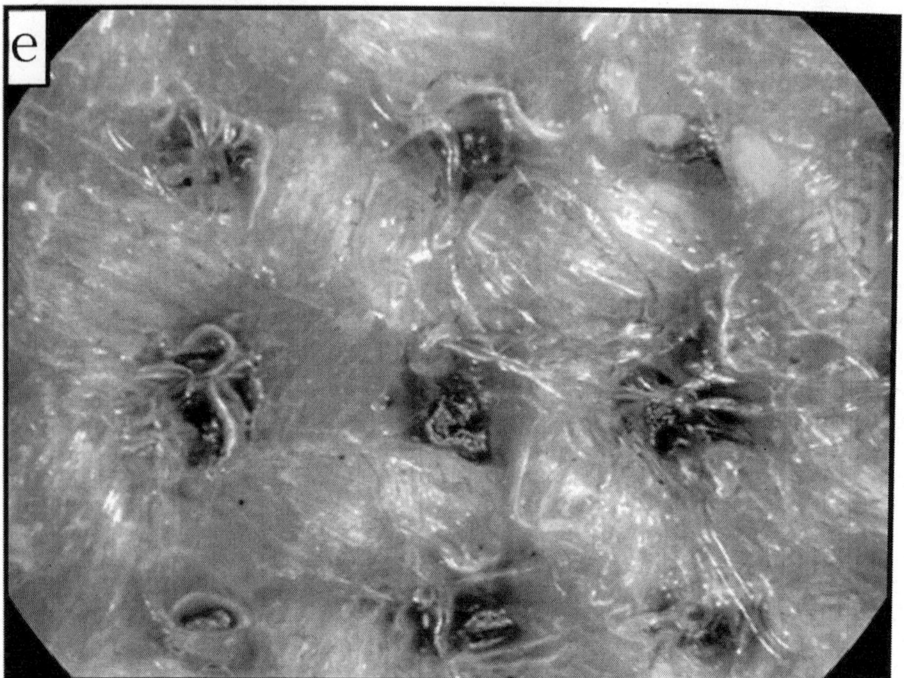

Figure 13.3. (e) The optical microscope images of nanofiber web after laminating at temperatures more than 140°C (at 100 magnification).

Also, to examine how laminating temperature affect the breathability of multilayer fabric, air permeability experiment was performed. Figure 13.4 indicates the effect of laminating temperature on air permeability. As might be expected, air permeability decreased with increasing laminating temperature. This behavior attributed to melting procedure of adhesive layer. As mentioned above, before melting point the adhesive gradually spreads on web surface. This phenomenon causes that the adhesive layer act like an impervious barrier to air flow and reduces air permeability of multilayer fabric. But at melting point and above, the penetration of melt adhesive into nanofiber/fabric structure leads to fill its pores and finally decrease in air permeability.

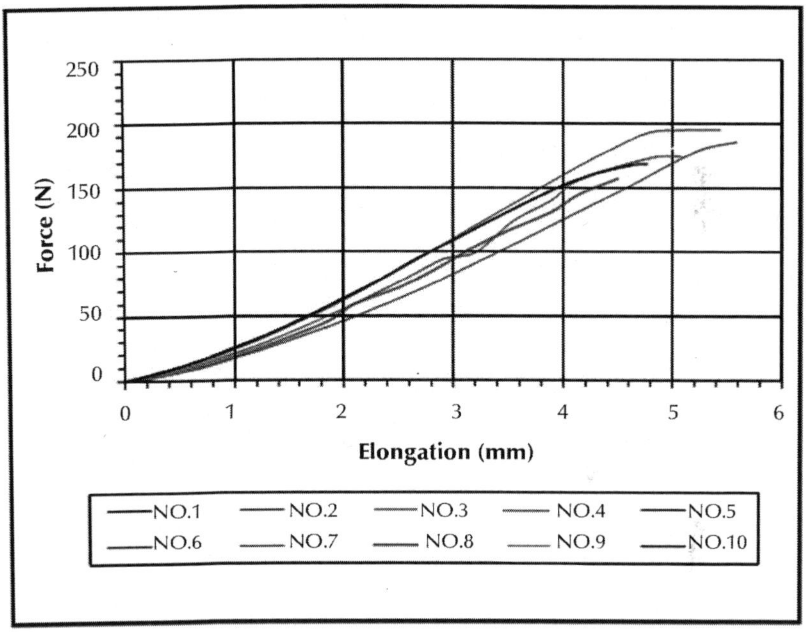

Figure 13.4. Air permeability of multilayer fabric as a function of laminating temperature.

Furthermore, we only observed that the adhesive force between layers was increased according to temperature rise. The Sample (a) exhibited very poor adhesion between nanofiber web and fabric and it could be separated by light abrasion of thumb, while adhesion increased by increasing laminating temperature to melting point. It must to be noted that after melting point because of passing of melt PPSN across nanofiber web, adhesion between two layers of fabric will occurred.

Mechanical Properties of Multilayer Nano-web

The tensile strength of multilayer fabrics with and without nanofiber web, were carried out using MICRO250 tensile machine (SDL International Ltd.). Ten samples were cut from the warp directions of multilayer fabric at size of 10mm×200mm and then exposed to the standard condition (25°C,60% RH) for 24h in order to conditioning.

To measure tensile strength, testing was performed by load cell of 25 Kgf. Also, the distance between the jaws and the rate of extension were selected 100 mm and 20 mm/min, respectively .The tensile strength of samples without nanofibers (Fig. 13.5) are weaker than those laminated with nanofibers (Fig. 13.6). According to Table 13.2, the breaking load and breaking elongation for the samples laminated with electrospun nanofibers are improved as well. These variations can be observed clearly in Figs. 13.7 and 13.8 for 10 samples.

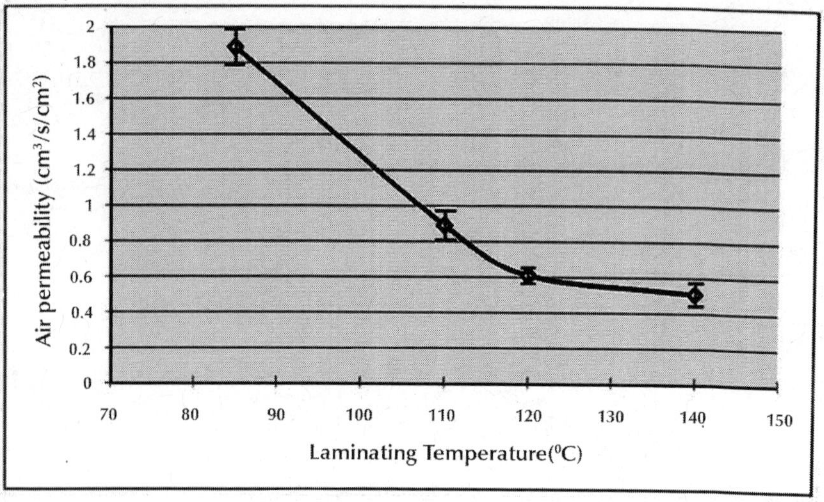

Figure 13.5. Force-elongation curve for multilayer fabric without nanofiber web.

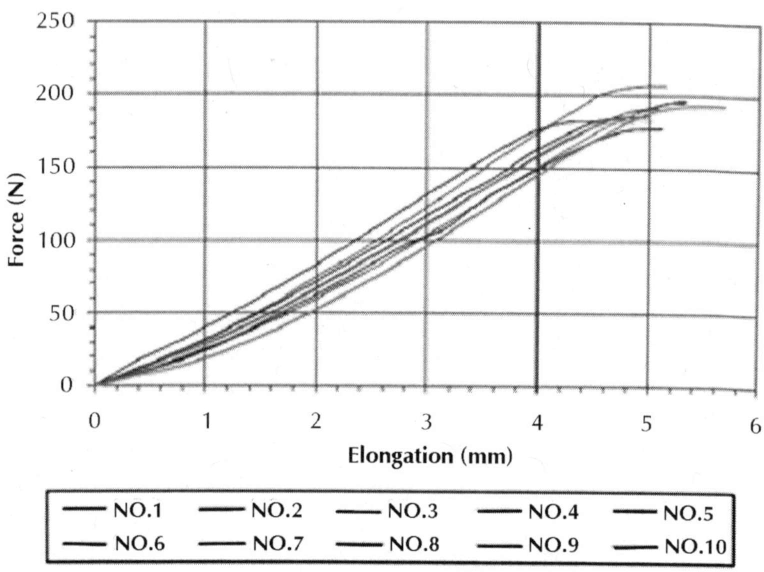

Figure 13.6. Force-elongation curve for multilayer fabric with nanofiber web.

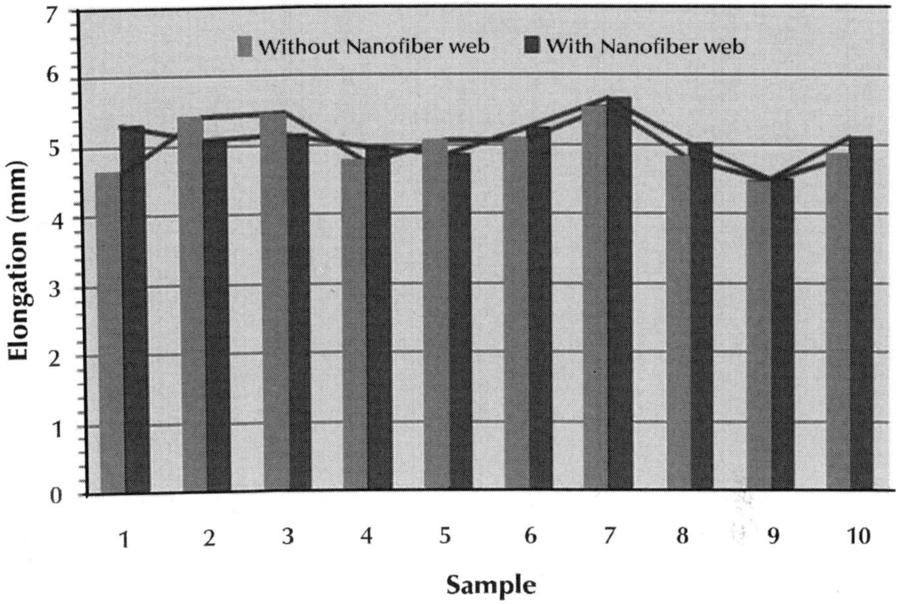

Figure 13.7. Breaking elongation of 10 samples.

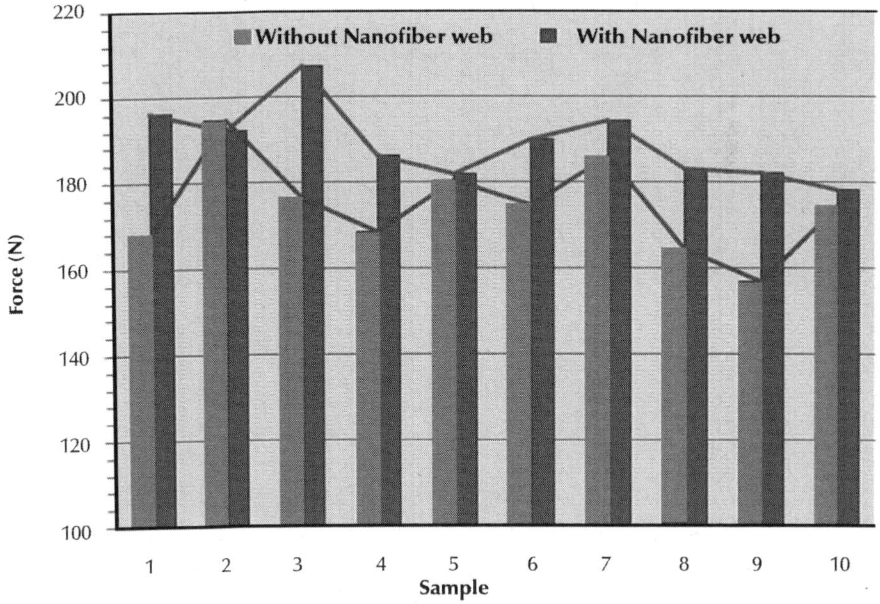

Figure 13.8. Breaking load of 10 samples.

Table 13.2. Processing parameters in electrospinning.

Solution properties	Viscosity
	Polymer concentration
	Molecular weight of polymer
	Electrical conductivity
	Elasticity
	Surface tension
Processing conditions	Applied voltage
	Distance from needle to collector
	Volume feed rate
	Needle diameter
Ambient conditions	Temperature
	Humidity
	Atmospheric pressure

Simulation of Nano-web

For the continuous fibers, it is assumed that the lines are infinitely long so that in the image plane, all lines intersect the boundaries. Under this scheme (Fig. 13.9), a line with a specified thickness is defined by the perpendicular distance d from a fixed reference point O located in the center of the image and the angular position of the perpendicular α. Distance d is limited to the diagonal of the image.

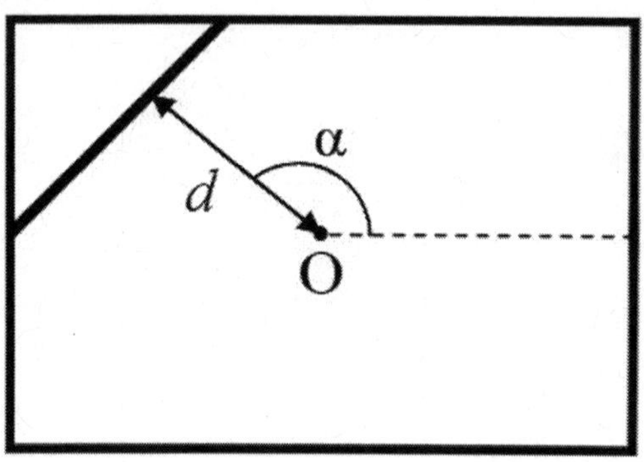

Figure 13.9. Procedure for μ-randomness.

Based on the objective of this chapter, several variables are allowed to be controlled during the simulation:

1. Web density that can be controlled using the line density which is the number of lines to be generated in the image.
2. Angular density which is useful for generating fibrous structures with specific orientation distribution. The orientation may be sampled from either a normal or a uniform random distribution.

3. Distance from the reference point normally varies between zero and the diagonal of the image, restricted by the boundary of the image and is sampled from a uniform random distribution.
4. Line thickness (fiber diameter) is sampled from a normal distribution. The mean diameter and its standard deviation are needed.
5. Image size can also be chosen as required.

Fiber Diameter Measurement

The first step in determining fiber diameter is to produce a high quality image of the web, called micrograph, at a suitable magnification using electron microscopy techniques. The methods for measuring electrospun fiber diameter are described in following sections.

MANUAL METHOD

The conventional method of measuring the fiber diameter of electrospun webs is to analyze the micrograph manually. The manual analysis usually consists determining the length of a pixel of the image (setting the scale), identifying the edges of the fibers in the image and counting the number of pixels between two edges of the fiber (the measurements are made perpendicular to the direction of fiber-axis), converting the number of pixels to *nm* using the scale and recording the result. Typically 100 measurements are carried out (Fig. 13.10).

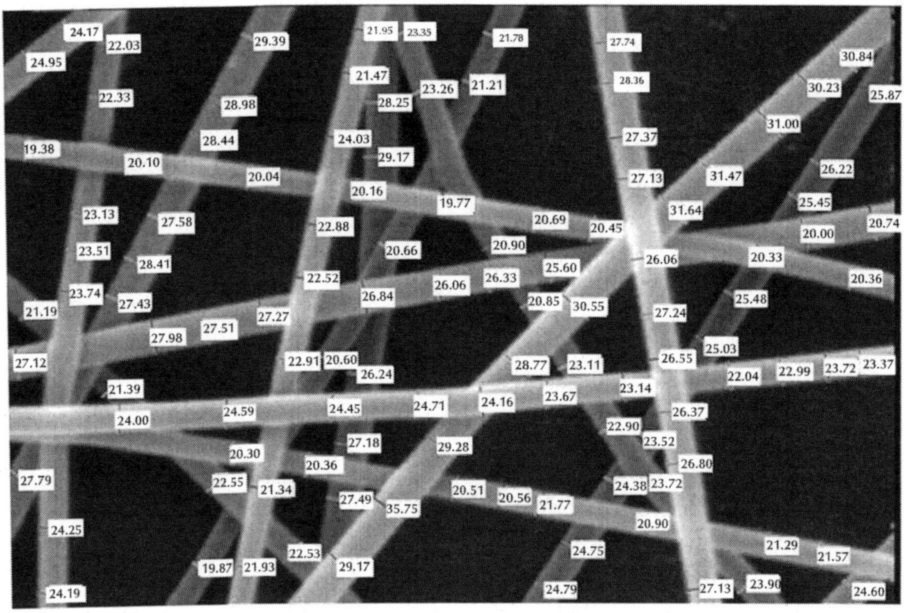

Figure 13.10. Manual method.

However, this process is tedious and time-consuming especially for large number of samples. Furthermore, it cannot be used as on-line method for quality control since an operator is needed for performing the measurements. Thus, developing automated techniques which eliminate the use of operator and has the capability of being employed as on-line quality control is of great importance.

Distance Transform

The distance transform of a binary image is the distance from every pixel to the nearest nonzero-valued pixel. The center of an object in the distance transformed image will have the highest value and lie exactly over the object's skeleton. The skeleton of the object can be obtained by the process of skeletonization or thinning. The algorithm removes pixels on the boundaries of objects but does not allow objects to break apart. This reduces a thick object to its corresponding object with one pixel width. Skeletonization or thinning often produces short spurs which can be cleaned up automatically with a pruning procedure.

The algorithm for determining fiber diameter uses a binary input image and creates its skeleton and distance transformed image. The skeleton acts as a guide for tracking the distance transformed image by recording the intensities to compute the diameter at all points along the skeleton. Figure 13.11 shows a simple simulated image, which consists of five fibers with diameters of 10, 13, 16, 19, and 21 pixels, together with its skeleton and distance map including the histogram of fiber diameter obtained by this method.

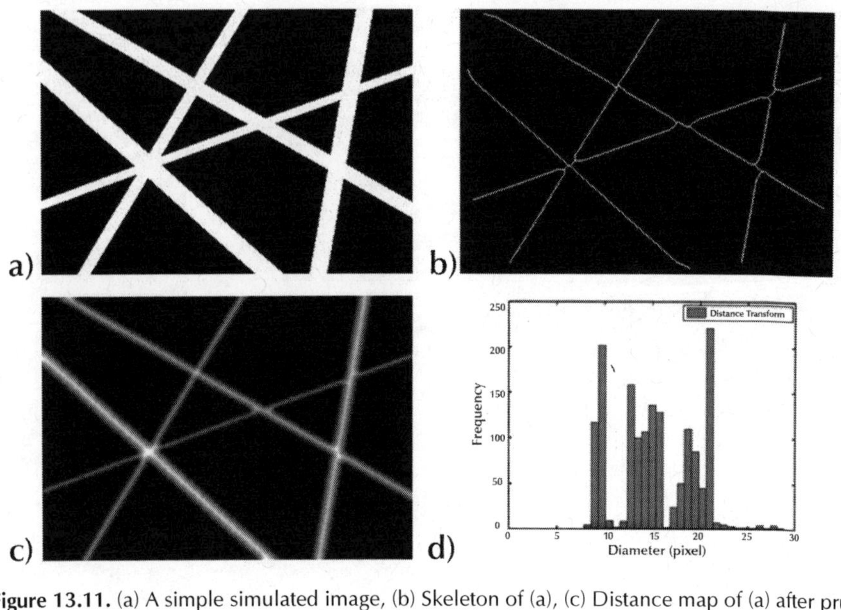

Figure 13.11. (a) A simple simulated image, (b) Skeleton of (a), (c) Distance map of (a) after pruning, (d) Histogram of fiber diameter distribution obtained by distance transform method.

Direct Tracking

Direct tracking method uses a binary image as an input data to determines fiber diameter based on information acquired from two scans; first a horizontal and then a vertical scan. In the horizontal scan, the algorithm searches for the first white pixel adjacent to a black. Pixels are counted until reaching the first black. The second scan is then started from the mid point of horizontal scan and pixels are counted until the first black is encountered. Direction changes if the black pixel is not found. Having the number of horizontal and vertical scans, the number of pixels in perpendicular direction which is the fiber diameter could be measured from a geometrical relationship. The explained process is illustrated in Fig. 13.12.

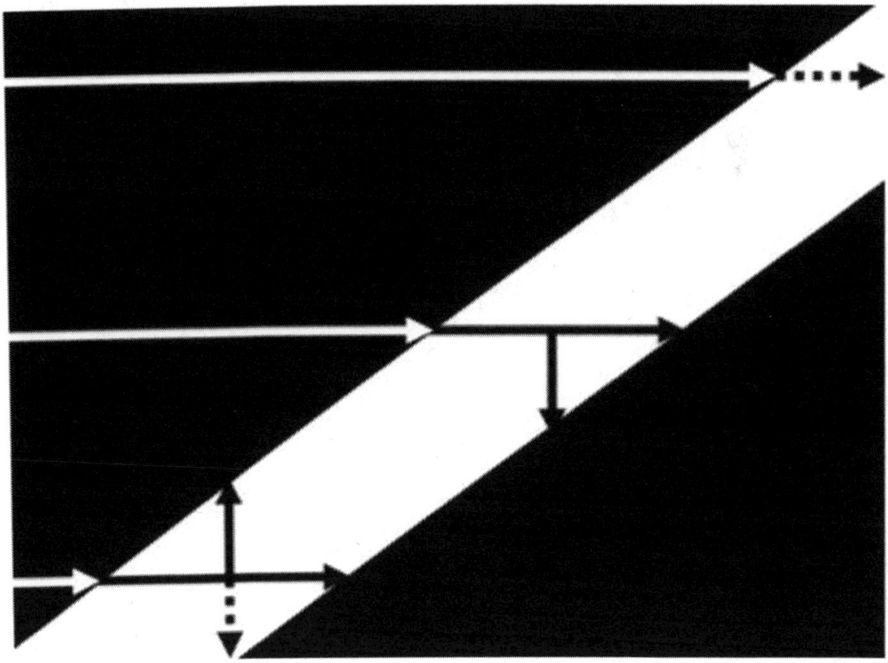

Figure 13.12. Diameter measurement based on two scans in direct tracking method.

In electrospun nonwoven webs, nanofibers cross each other at intersection points and this brings about the possibility for some untrue measurements of fiber diameter in these regions. To circumvent this problem, a process called fiber identification is employed. First, black regions are labeled and couple of regions between which a fiber exists is selected. In the next step, the two selected regions are connected performing a dilation operation with a large enough structuring element. Dilation is an operation that grows or thickens objects in a binary image by adding pixels to the boundaries of objects. The specific manner and extent of this thickening is controlled by the size and shape of the structuring element used. In the following process, an erosion operation with the same structuring element is performed and the fiber is recognized. Erosion shrinks or thins objects in a binary image by removing pixels on object boundaries. As in dilation, the manner and extent of shrinking is controlled by a structuring element. Then, in order to enhance the processing speed, the image is cropped to the size of selected regions. Afterwards, fiber diameter is measured according to the previously explained algorithm. This trend is continued until all of the fibers are analyzed. Finally, the data in pixels may be converted to *nm* and the histogram of fiber diameter distribution is plotted. Fig. 13.13 shows a labeled simple simulated image and the histogram of fiber diameter obtained by this method.

a)
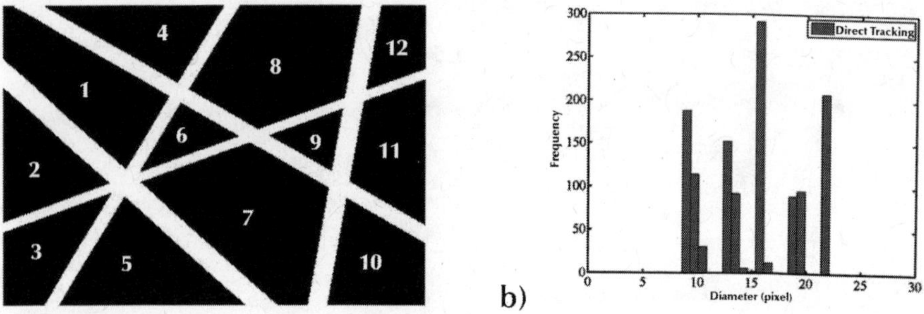

Figure 13.13. (a) A simple simulated image which is labeled, (b) Histogram of fiber diameter distribution obtained by direct tracking.

KEYWORDS

- **Angular density**
- **Direct tracking**
- **Electrospinning**
- **Multilayer fabrics**

Chapter 14

Update on Control of Electrospun Nanofiber Diameter—Part III

INTRODUCTION

In electrospinning process, a high electric field is generated between a polymer solution held by its surface tension at the end of a syringe (or a capillary tube) and a collection target.

In the fabric lamination, producing an adhesive bond which guarantees no delaminating or failure in use requires lamination skills and information about adhesive types. It is relatively simple to create a strong bond; the challenge is to preserve the original properties of the fabric and to produce a flexible laminate with the required appearance, handle and durability. In other words, the application of adhesive should have minimum affect on the fabric flexibility and aesthetics during the lamination process 63, therefore, adhesive must be applying in a controlled manner. In order to achieve to this purpose, it is generally necessary that the least amount of a highly effective adhesive applied and it penetrate to a certain extent of the fabric and cover the widest possible surface area. Too much adhesive and excessive penetration likely to lead to fabric stiffening and it could result in thermal discomfort in the cloth; since the adhesive itself could form an impermeable barrier to perspiration.

The adhesives could be as solvent/water-based adhesive or as hot-melt adhesive. In first group, the adhesives are as solutions in solvent or water, and solidify by evaporating of the carrying liquid. In this group, Solvent-based adhesives could "wet" the surfaces to be joined better than water-based adhesives, and also could solidify faster. But unfortunately, they are environmentally unfriendly, usually flammable, and more expensive than those. Of course it is not means that the water-based adhesives are always preferred to laminating, since in practice, drying off water in terms of energy and time is expensive too. Beside, water-based adhesives are not resisting to water or moisture because of their hydrophilic nature. But in hot-melt adhesive group, the adhesives are as solids and melt under the action of heat. These types of adhesives are environmentally friendly, inexpensive, require less heat and energy, and so is now more preferred. They can be of several different chemical types, such as polyolefin (polyethylene, polypropylene), polyurethane, polyester, polyamide or blends of different polymers or copolymers in order to reach for a wide range of properties (including melting points, durability to washing and dry cleaning, and heat resistance). Hot-melt lamination can be either continuous (hot calendars) or static (flat iron or Hoffman press) and is accomplished by two separate processes: first a means of applying the actual adhesive; and second bringing the two substrates together to form the actual bond under the action of heat and pressure. In this process, the heating accomplish at temperatures above the softening or melting point of adhesive. In addition, hot melt adhesives are available in several forms; as a web, as a continuous film, or in powder or granular form. The adhesive powders are available in most chemical types and also in particle sizes ranging from very small up to about 500 μm or so in diameter. Adhesives in film or web form are more expensive than the corresponding adhesive powders. The webs are discontinuous and produce laminates which are flexible, porous, and breathable, whereas, Continuous film adhesives cause stiffening and produce laminates which are not porous and permeable to both air and water vapor. This behavior attributed to impervious nature of adhesive film and its shrinkage under the action of heat.

Figure 14.1 represents the optical microscope image of multilayer nanofiber web. Accurate and automated measurement of nanofiber diameter of laminated webs is useful and crucial and therefore has been taken into consideration in this contribution. The objective of the current research would then be to develop an image analysis based method to serve as a simple, automated and efficient alternative for electrospun nanofiber diameter measurement with particular application in laminated nanofiber web.

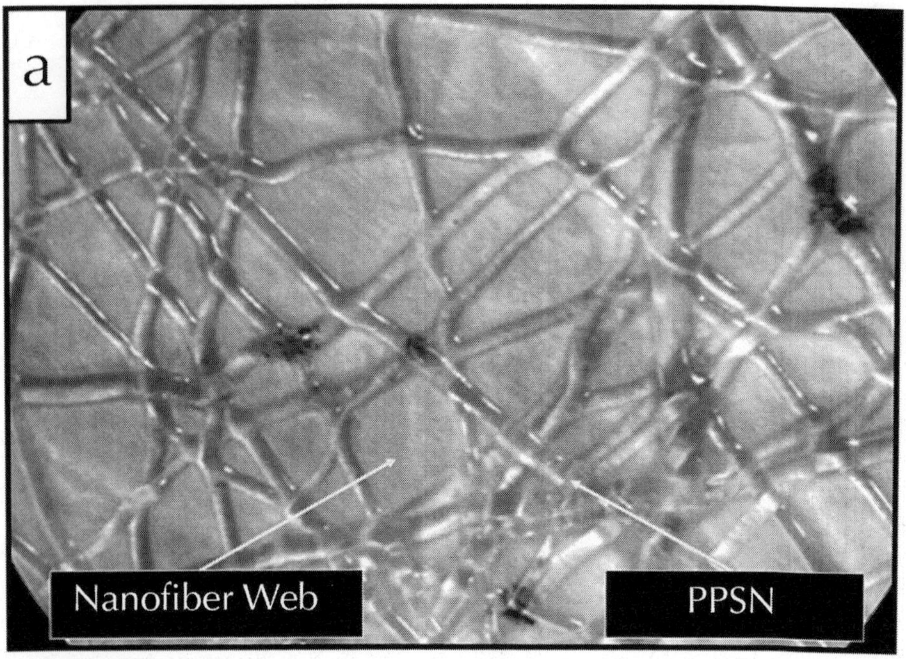

Figure 14.1. The optical microscope images of multilayer nanofiber web (PPSN):polypropylene spun-bond nonwoven.

METHODOLOGY

The algorithm for determining fiber diameter uses a binary input image and creates its skeleton and distance transformed image (distance map). The skeleton acts as a guide for tracking the distance transformed image and fiber diameters are measured from the intensities of the distance map at all points along the skeleton. Figure 14.2 shows a simple simulated image, which consists of five fibers with diameters of 10, 13, 16, 19, and 21 pixels, together with its skeleton and distance map including the histogram of fiber diameter obtained by this method.

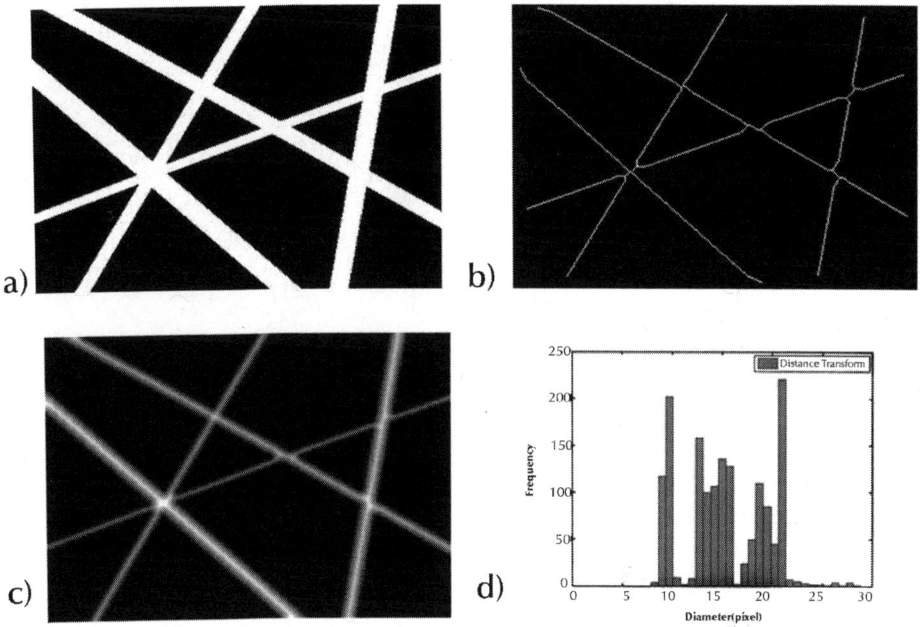

Figure 14.2. (a) A simple simulated image, (b) Skeleton of (a), (c) Distance map of (a) after pruning, (d) Histogram of fiber diameter distribution obtained by distance transform method.

In this chapter, we developed direct tracking method for measuring electrospun nanofiber diameter. This method which also uses a binary image as the input, determines fiber diameter based on information acquired from two scans; first a horizontal and then a vertical scan. In the horizontal scan, the algorithm searches for the first white pixel (representative of fibers) adjacent to a black (representative of background). Pixels are then counted until reaching the first black. Afterwards, the second scan is started from the midpoint of horizontal scan and pixels are counted until the first vertical black pixel is encountered. Direction will change if the black pixel isn't found (Fig. 14.3). Having the number of horizontal and vertical scans, the number of pixels in perpendicular direction which is the fiber diameter in terms of pixels can be measured through a simple geometrical relationship.

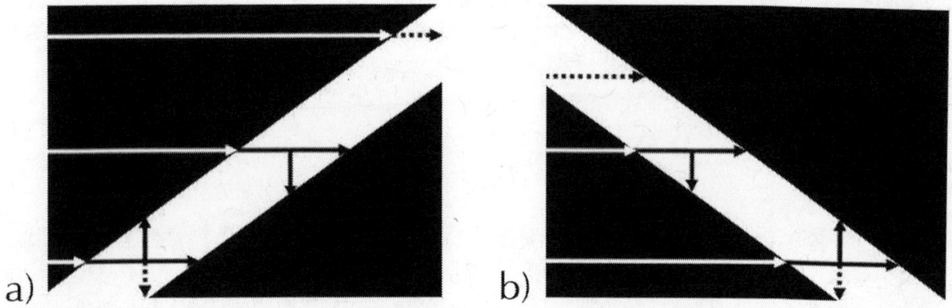

Figure 14.3. Fiber diameter measurement based on two scans in direct tracking method.

In electrospun webs, nanofibers cross each other at intersection points and this brings about the possibility for some untrue measurements of fiber diameter in these regions. To circumvent this problem, a process called fiber identification is employed. First, black regions are labeled and a couple of regions between which a fiber exists, are selected. Figure 14.4 depicts the labeled simulated image and the histogram of fiber diameter obtained by direct tracking method.

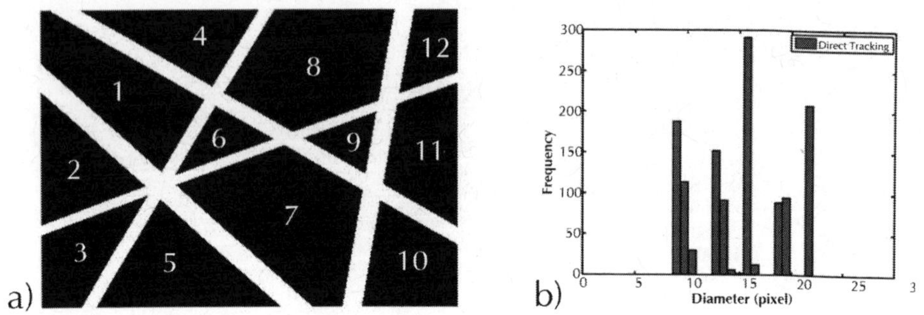

Figure 14.4. (a) The labeled simulated image, (b) Histogram of fiber diameter distribution obtained by direct tracking method.

Now, reliable evaluation of the accuracy of the developed methods requires samples with known characteristics. Since it is neither possible to obtain real electrospun webs with specific characteristics through the experiment nor there is a method which measures fiber diameters precisely with which to compare the results, the method will not be well evaluated using just real webs. To that end, a simulation algorithm has been employed for generating samples with known characteristics. In this case, it is assumed that the lines are infinitely long so that in the image plane, they intersect the boundaries. Under this scheme, which is shown in Fig. 14.5, a line with a specified thickness is defined by the perpendicular distance d away from a fixed reference point O located in the center of the image and the angular position of the perpendicular α. Distance d is limited to the diagonal of the image. Several variables are allowed to be controlled during simulation; line thickness, line density, angular density, and distance from the reference point. These variables can be sampled from given distributions or held constant.

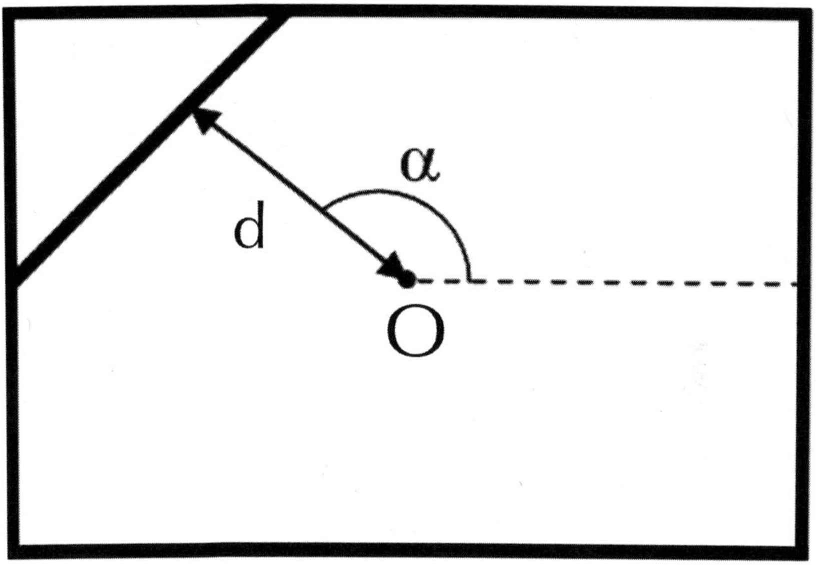

Figure 14.5. μ-randomness procedure.

Distance transform and direct tracking algorithms for measuring fiber diameter both require binary image as their input. Hence, the micrographs of electrospun webs first have to be converted to black and white. This is carried out by thresholding process (known also as segmentation) which produces binary image from a grayscale (intensity) image. This is a critical step because the segmentation affects the result significantly. Prior to the segmentation, an intensity adjustment operation and a two dimensional median filter are often applied in order to enhance the contrast of the image and remove noise.

In the simplest thresholding technique, called global thresholding, the image is segmented using a single constant threshold. One simple way to choose a threshold is by trial and error. Each pixel is then labeled as object or background depending on whether its gray level is greater or less than the value of threshold respectively.

The main problem of global thresholding is its possible failure in the presence of non-uniform illumination or local gray level unevenness. An alternative to this problem is to use local thresholding instead. In this approach, the original image is divided to subimages and different thresholds are used for segmentation. As it is shown in Fig. 14.6, global thresholding resulted in some broken fiber segments. This problem was solved using local thresholding.

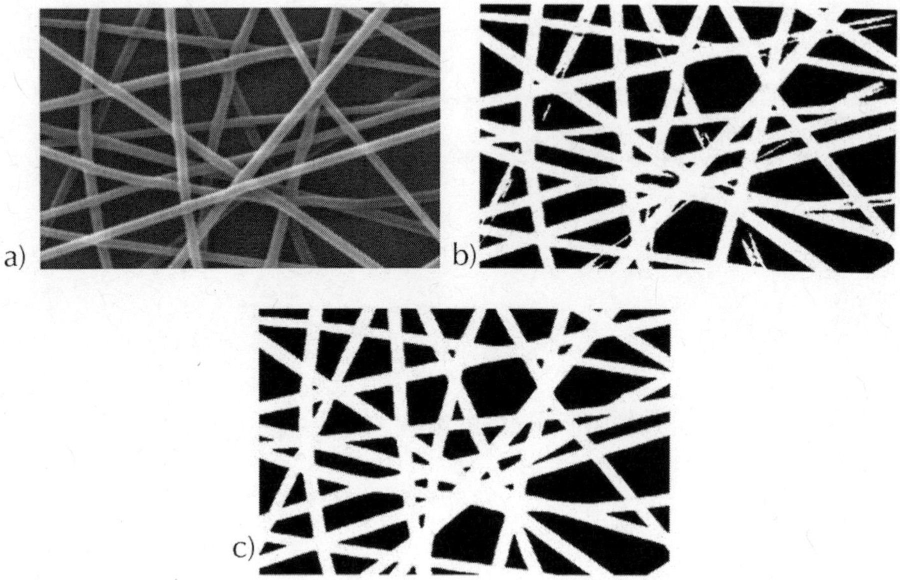

Figure 14.6. (a) A typical electrospun web, (b) Global thresholding, (c) Local thresholding.

EXPERIMENT

Electrospun nanofiber webs used as real webs in image analysis were prepared by electrospinning aqueous solutions of PVA with average molecular weight of 72,000 g/mol (MERCK) at different processing parameters. The micrographs of the webs were obtained using a Philips (XL-30) environmental Scanning Electron Microscope (SEM) under magnification of 10,000X after gold sputter coating.

RESULTS AND DISCUSSION

Three simulated images generated by μ-randomness procedure were used as samples with known characteristics to demonstrate the validity of the techniques. They were each produced by 30 randomly oriented lines with varied diameters sampled from normal distributions with mean of 15 pixels and standard deviation of 2, 4, and 8 pixels respectively. Table 14.1 summarizes the structural features of these simulated images which are shown in Fig. 14.7.

Table 14.1. Structural characteristics of the simulated images generated using μ-randomness procedure.

No.	Angular range	Line density	Line thickness	
			M	Std
1	0–360	30	15	2
2	0–360	30	15	4
3	0–360	30	15	8

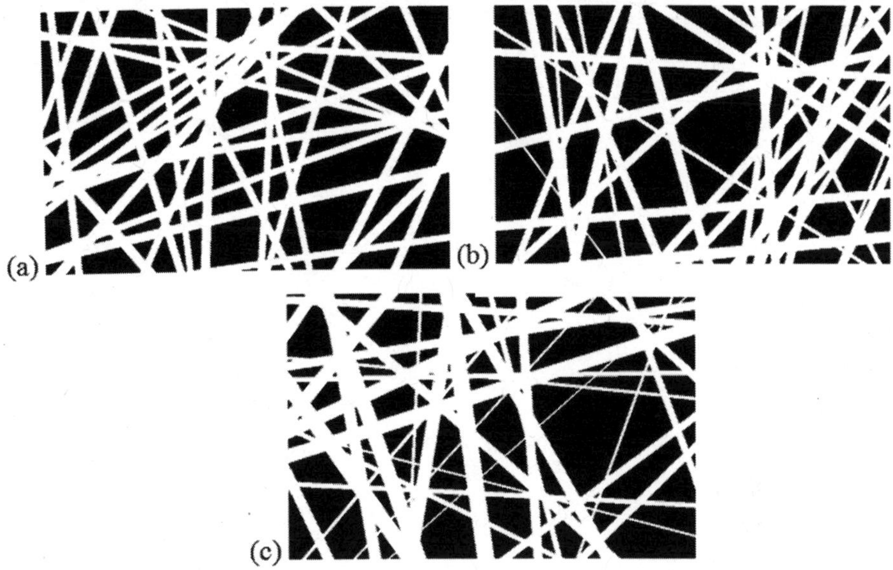

Figure 14.7. Simulated images generated using μ-randomness procedure.

Mean and standard deviation of fiber diameters for the simulated images obtained by direct tracking as well as distance transform are listed in Table 14.2. Figure 14.8 shows histograms of fiber diameter distribution for the simulated images obtained by the two methods. In order to make a true comparison, the original distribution of fiber diameter in each simulated image is also included. The line over each histogram is related to the fitted normal distribution to the corresponding fiber diameters.

Table 14.2. Mean and standard deviation of fiber diameters for the simulated images.

		No. 1	**No. 2**	**No. 3**
Simulation	M	15.247	15.350	15.367
	Std	1.998	4.466	8.129
Distance transform	M	16.517	16.593	17.865
	Std	5.350	6.165	9.553
Direct tracking	M	16.075	15.803	16.770
	Std	2.606	5.007	9.319

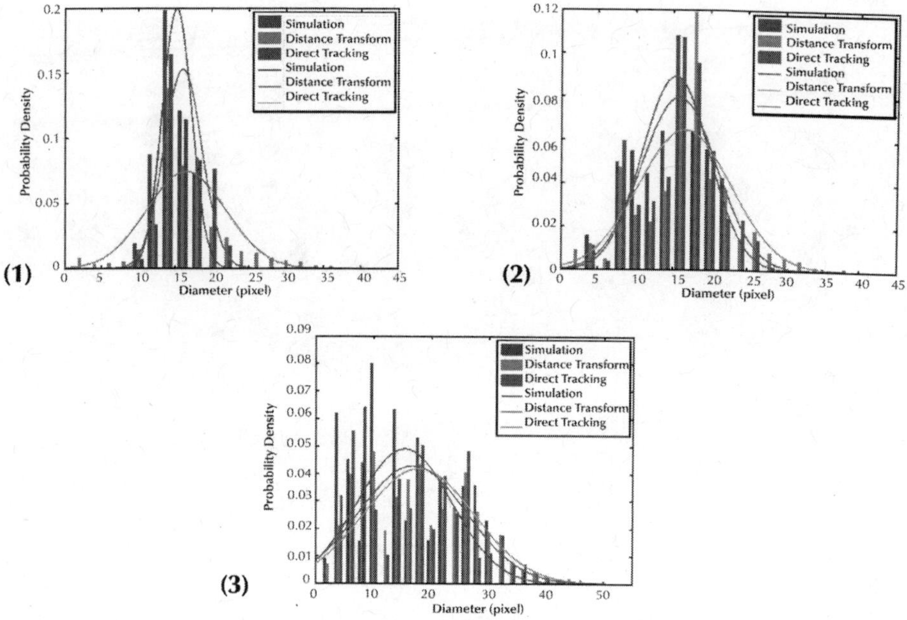

Figure 14.8. Histograms of fiber diameter distribution for the simulated images.

Table 14.2 and Fig. 14.8 clearly demonstrate that for all simulated webs, direct tracking method resulted in mean and standard deviation of fiber diameters which are closer to those of the corresponding simulated image (the true ones). Distance transform method is far away from making reliable and accurate measurements. This may be due to remaining some branches in the skeleton even after pruning. Thicker the line, higher the possibility of branching during skeletonization (or thinning). Although these branches are small, their orientation is typically normal to the fiber axis; thus causing widening the distribution obtained by distance transform method.

KEYWORDS

- **Electrospun webs**
- **Global thresholding**
- **Hot-melt lamination**
- **Solvent/water-based adhesive**

References

Haghi, A. K. and Akbari, M. (2007). *Physica Status Solidi* **204**, 1830.

Kanafchian, M., Valizadeh, M., and Haghi, A. K. (2011). *Korean Journal of Chemical Engineering* **28**, 428.

Kanafchian, M., Valizadeh, M., and Haghi, A. K. (2011). *Korean Journal of Chemical Engineering* **28**, 445.

Kanafchian, M., Valizadeh, M., and Haghi, A. K. (2011). *Korean Journal of Chemical Engineering* **28**, 751.

Kanafchian, M., Valizadeh, M., and Haghi, A. K. (2011). *Korean Journal of Chemical Engineering* **28**, 763.

Mottaghitalab, V. and Haghi, A. K. (2011). *Korean Journal of Chemical Engineering* **28**, 114.

Ziabari, M., Mottaghitalab, V., and Haghi, A. K. (2008). *Korean Journal of Chemical Engineering* **25**, 919.

Ziabari, M., Mottaghitalab, V., and Haghi, A. K. (2008). *Korean Journal of Chemical Engineering* **25**, 923.

Ziabari, M., Mottaghitalab, V., and Haghi, A. K. (2009). *Brazilian Journal of Chemical Engineering* **26**, 53.

Ziabari, M., Mottaghitalab, V., and Haghi, A. K. (2010). *Korean Journal of Chemical Engineering* **27**, 340.

Ziabari, M., Mottaghitalab, V., McGovern, S. T., and Haghi, A. K. (2007). *Nanoscale Research Letter* **2**, 297.

Ziabari, M., Mottaghitalab, V., McGovern, S. T., and Haghi, A. K. (2008). *Chinese Physics Letters* **25**, 3071.

Index

I

M

N

O

P

R

S

T